Peter D Riley

Cambridge
checkpoint

ENDORSED BY

CAMBRIDGE
International Examinations

NEW EDITION

checkpoint
Science

1

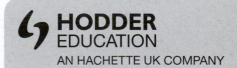

HODDER
EDUCATION
AN HACHETTE UK COMPANY

Titles in this series

Cambridge Checkpoint Science Student's Book 1	978 1444 12603 7
Cambridge Checkpoint Science Teacher's Resource Book 1	978 1444 14380 5
Cambridge Checkpoint Science Student's Book 2	978 1444 14375 1
Cambridge Checkpoint Science Teacher's Resource Book 2	978 1444 14381 2
Cambridge Checkpoint Science Student's Book 3	978 1444 14378 2
Cambridge Checkpoint Science Teacher's Resource Book 3	978 1444 14382 9

To Stephen and Margaret

Hachette UK's policy is to use papers that are natural, renewable and recyclable products and made from wood grown in sustainable forests. The logging and manufacturing processes are expected to conform to the environmental regulations of the country of origin.

Orders: please contact Bookpoint Ltd, 130 Milton Park, Abingdon, Oxon OX14 4SB. Telephone: (44) 01235 827720. Fax: (44) 01235 400454. Lines are open 9.00–5.00, Monday to Saturday, with a 24-hour message answering service. Visit our website at www.hoddereducation.co.uk

© Peter D Riley 2005
First published in 2005 by
Hodder Education, an Hachette UK Company,
Carmelite House, 50 Victoria Embankment,
London EC4Y 0DZ

This second edition first published 2011

Impression number   17
Year   2018

Illustrations by Barking Dog and Pantek Arts Ltd
Typeset in 12/14pt ITC Garamond Light by Pantek Media, Maidstone, Kent
Printed in India

A catalogue record for this title is available from the British Library

ISBN 978 1444 12603 7

Contents

Preface

To the student

Science is the study of everything in the universe and scientists go about their studies in a very special way. They use the scientific method, which is about making a scientific enquiry or investigation. It helps scientists build up knowledge – scientific knowledge – about how things are and how things happen.

Science, then, has two parts. One part is about making a scientific enquiry and this process involves having ideas, making observations and carrying out investigations. The other part is the huge collection of scientific facts – from the colour of a butterfly on a tree or the shape of a galaxy in space to what happens when we breathe and how a volcano can suddenly blow its top! In this book, we are going to look at how *you* can make scientific enquiries too, and how *you* can also learn about the many facts that scientists have discovered.

Cambridge Checkpoint Science covers the requirements of your examinations in a way that I hope will help you understand how observations, investigations and ideas have led to the scientific facts we use today. The questions are set to help you extract information from what you read and see, and to help you think more deeply about each chapter in the book. Some questions are set so you can discuss your ideas with others and sometimes develop a point of view on different scientific issues. This should help you in the future when new scientific issues, which are as yet unknown, affect your life.

The scientific activities of thinking up ideas to test and carrying out investigations are enjoyed so much by many people that they take up a career in science. Perhaps *Checkpoint Science 1* might help you to take the first step towards a career in science too.

To the teacher

Checkpoint Science 1 has been developed from *Checkpoint Biology*, *Checkpoint Chemistry* and *Checkpoint Physics* to cover the requirements of the Cambridge Checkpoint Tests and other equivalent junior secondary science courses. It also has three further aims:

- to help students become more scientifically literate by encouraging them to examine the information in the text and illustrations in order to answer questions about it in a variety of ways
- to encourage students to talk together about what they have read
- to present science as a human activity by considering the development of scientific ideas from the earliest times to the present day.

The student's book begins with a chapter called *Introducing science* where the separate sciences of biology, chemistry and physics are presented in the context of the work of present day scientists. Items of general laboratory apparatus, including the Bunsen burner and spirit burner, are introduced before the requirements for Scientific Enquiry are set out for stage 7 of the Cambridge Secondary 1 Science Curriculum Framework. This is followed by a feature on the history of the development of scientific enquiry, and then the students are set tasks that are involved in carrying out investigations. The chapter ends by looking at safety in the laboratory.

The chapters that follow are arranged in sections with Chapters 1–7 addressing the learning requirements for biology stage 7, Chapters 8–12 addressing the learning requirements for chemistry stage 7 and Chapters 13–17 addressing the learning requirements for physics stage 7 of the Cambridge Secondary 1 Science Curriculum Framework.

The student's book is supported by a teacher's resource book that provides answers to all the questions in the student's book – those in the body of the chapter and those that occur as end of chapter questions. Each chapter is supported by a chapter in the teacher's resource book which features a summary, chapter notes providing additional information and suggestions, a curriculum framework reference table, practical activities (some of which can be used for assessing Scientific Enquiry skills), homework activities, a 'lesson ideas' section integrating the practical activities and homework activities, and two Practice Tests with marking guidance.

Peter D Riley
May 2011

Acknowledgements

The author would like to thank Ian Lodge for reading and advising on early stages of the manuscript.

The Publishers would like to thank the following for permission to reproduce copyright material:

Photo credits

Cover © Steven Miric/Photodisc/Getty Images; **p.1** *l* © Digital Vision/Getty Images; *r* © Photodisc/Getty Images; *c* © Getty Images/Image Source; **p.2** *t* © Richard Carey – Fotolia; *c* © Sipa Press/Rex Features; *b* © Frank Zullo/Science Photo Library; **p.5** *tl* © Andrew Lambert/Science Photo Library; *tc* © Andrew Lambert/Science Photo Library; *tr* © Andrew Lambert/Science Photo Library; *cl* © Andrew Lambert/Science Photo Library; *cr* © Andrew Lambert/Science Photo Library; *bl* © Andrew Lambert/Science Photo Library; *br* © Andrew Lambert/Science Photo Library; **p10** © David Lyons/Alamy; **p.11** © Ariadne Van Zandbergen/Alamy; **p.12** © Health Protection Agency/Science Photo Library; **p.15** *c* © Alexander Kosarev – Fotolia; *r* © Eric Etman – Fotolia; **p.20** © Peter Chadwick/Science Photo Library; **p.21** © Claude Nuridsany & Marie Perennou /Science Photo Library; **p.22** © Corey Hochachka/Design Pics Inc./Rex Features; **p.23** © Martin Dohrn/Science Photo Library; **p.24** © Anatol – Fotolia; **p.25** © Robert Hardholt – Fotolia; **p.27** *t* © NASA/Science Photo Library; *b* OAR/National Undersea Research Program (NURP); National Oceanic and Atmospheric Administration/Department of Commerce; **p.29** © KPA/Zuma/Rex Features; **p.31** *t* © Geoff Tompkinson/Science Photo Library; *b* © hazel proudlove – Fotolia; **p.32** © Ina Raschke – Fotolia; **p.37** © Natalya Antoshchenko – Fotolia; **p.43** Wikimedia public domain/http:// commons.wikimedia.org/wiki/File:Vesalius_Fabrica_p174.jpg; **p.45** © The Art Gallery Collection/Alamy; **p.46** Wikimedia public domain /http:// commons.wikimedia.org/wiki/File:RobertHookeMicrographia1665.jpg; **p.49** © Astrid & Hanns-Frieder Michler/Science, Photo Library; **p.55** © Sinclair Stammers/Science Photo Library; **p.57** © Natural Visions/Alamy; **p.58** *t* © Steve Gschmeissner/Science Photo Library; *cl* © Dick Makin/ Last Resort Photo Library; *cr* © Dick Makin/Last Resort Photo Library; *bl* © Dick Makin/Last Resort Photo Library; *br* © Dick Makin/Last Resort Photo Library; **p.59** © Scimat/Science Photo Library; **p.62** © Graham Corney/Alamy; **p.63** © North Wind Picture Archives/Alamy; **p.67** © Dr Jeremy Burgess/Science Photo Library; **p.69** © John Cancalosi/Alamy; **p.70** ©Sally Morgan/Ecoscene; **p.71** © Jacomina Wakeford/ICCE; **p.76** © hazel proudlove – Fotolia; **p.77** © Renaud Visage/Photographer's Choice/Getty Images; **p.78** © Imagestate Media; **p.79** © Imagestate Media; **p.80** *tl* © John Devries/Science Photo Library; *tr* Per Harald Olsen/http://creativecommons.org/licenses/by/2.5/deed.en; *b* © Tom Mchugh/Science Photo Library; **p.82** *l* © Imagestate Media; *r* © Julien Méjean – Fotolia; **p.83** © Science Photo Library/Alamy; **p.84** © Danita Delimont/Alamy; **p.85** *t* © Stephen Dalton/NHPA/photoshot; *b* © Doug Allan/Science Photo Library; © Fred Bavendam/Still pictures/Specialist Stock; **p.87** © Imagestate Media; **p.88** © Imagestate Media; **p.91** © Neil Harris/Alamy; **p.92** *t* © Chris Lofty – Fotolia; *b* © Rene Wouters – Fotolia; **p.93** © Katja Xenikis – Fotolia; **p.94** *t* Library of Congress Prints & Photographs Division/LC-DIG-nclc-01336; *b* © Ian Bracegirdle/iStockphoto.com; **p.95** *t* © Hartmut Scwarzbach/Still Pics/Specialist Stock; *b* © manfredxy – Fotolia.com; **p.96** *t* © Jim West/Alamy; *b* © Colin Garratt; Milepost 92 ½/CORBIS; **p.97** *t* © Photodisc/Getty Images; *b* © David Pearson/Alamy; **p.98** © Phillip Wallick/Agstockusa/Science Photo Library; **p.99** *t* © Ecoscene/Wayne Lawler; *b* © Justin Sullivan/Getty Images; **p.101** © Natural Visions/Alamy; *b* NASA/Goddard Space Flight Center; **p.102** © Marc Schlossman/Panos Pictures; **p.103** © Theo Allofs/Corbis; **p.107** *t* © Vinicius Tupinamba – Fotolia; *b* © Microfield Scientific Ltd/Science Photo Library; **p.108** *l* © Michael & Patricia Fogden/CORBIS; *r* © Andrew J. Martinez/Science Photo Library; **p.109** *tl* © James L Davidson – Fotolia; *tr* © Gino Santa Maria – Fotolia; *b* © Georgette Douwma/Science Photo Library; **p.111** *t* © Kim Westerskov/Photographer's Choice RF/Getty Images; *c* © Richard Herrmann/ Photolibrary.com; *b* © Juniors Bildarchiv/Alamy; **p.112** *t* © The Art Gallery Collection/Alamy; *b* © Alessandro Mancini/Alamy; **p.113** *t* © imagebroker/Alamy; *l* © Phil McDermott/Alamy; *r* © Yuriy Kulik – Fotolia.com; **p.114** *t* © Imagestate Media; *c* © Imagestate Media; **p.115** *t* © Imagestate Media; *bl* ©Michael Ireland – Fotolia; *br* © hotshotsworldwide – Fotolia; **p.117** © Big Cheese Photo LLC/Alamy; **p.118** ©Antonio Nunes – Fotolia; **p.121** © David Levenson/Alamy; **p.123** © Imagestate Media; **p.125** *t* © Haydn Hansell/Alamy; *b* © Northwestern University/ Science Photo Library; **p.127** © Bob Battersby/BDI Images; **p.128** © Igor Groshev – Fotolia; **p.129** © Jim Nicholson/Alamy; **p.130** © cesco fotografo – Fotolia; **p.131** *all* © Andrew Lambert/Science Photo Library; **p.132** © David Parker & Julian Baum/Science Photo Library; **p.133** *l* © Martin Diebel/fStop/photolibrary.com; *r* © David Parker & Julian Baum/Science Photo Library; **p.134** © Bill Bachmann/Alamy; **p.135** *t* © Andrew Lambert Photography/Science Photo Library; *b* © beltsazar – Fotolia; **p.137** *t* © Iconotec/Alamy; *b* © The Art Gallery Collection/Alamy; **p.138** *t* © INTERFOTO/Alamy; *b* © Ben Mangor/SuperStock; **p.139** *l* © Lagui – Fotolia; *r* © Anyka – Fotolia; **p.140** © Jgz – Fotolia; **p.141** *t* © Ocean/ Corbis; *b* © Sipa Press/Rex Features; **p.142** *t* © Owaki – Kulla/CORBIS; *b* ©fotografiche.eu – Fotolia; **p.144** *b* © Imagestate Media; **p.146** © Science Photo Library; **p.147** *bl* © Imagestate Media; *t* © Gregory Dimijian/Science Photo Library; *bl* © illustrez-vous – Fotolia; **p.148** *t* © Andrew Lambert/ Science Photo Library; *br* © Andrew Lambert/Science Photo Library; *br* © Andrew Lambert/Science Photo Library; **p.149** © Martyn F. Chillmaid/ Science Photo Library; **p.150** *tl* © Mary Evans Picture Library 2010; **p.151** © The Carlsberg Archives; **p.152** © Dr Jeremy Burgess/Science Photo Library; **p.153** *l* © Martyn Chillmaid; *r* © Martyn Chillmaid; **p.154** *t* © Ecoscene/Chinch Gryniewicz; *b* © Mark Boulton/Alamy; **p.158** *t* NASA, ESA and AURA/Caltech; *b* © Jeff Hester And Paul Scowen, Arizona State University/Science Photo Library; **p.161** © Gavin Newman/Alamy; **p.163** *l* © michal812 – Fotolia; *r* ©Trevor Clifford Photography/Science Photo Library; **p.164** *tl* © Scientifica, Visuals Unlimited/Science Photo Library; *tr* © Finnbarr Webster/Alamy; *b* © Krafft/Hoa-Qui/Science Photo Library; **p.165** *t* U.S. Geological Survey; *b* © Scientifica, Visuals Unlimited/Science Photo Library; **p.166** *t* © Lyroky/Alamy; *c* © Scientifica, Visuals Unlimited/Science Photo Library; *bl* © Scientifica, Visuals Unlimited/Science Photo Library; *br* ©Tyler Boyes – Fotolia; **p.167** © Stephen Meese – Fotolia; **p.168** *l* © Yogesh More – Fotolia; *r* © John Morrison/Alamy; **p.169** *t* © The Travel Library/Rex Features; *bl* © ste72 – Fotolia; *br* © ibesed – Fotolia; **p.170** *l* © Leslie Garland Picture Library/Alamy; *r* © Martin Land/Science Photo Library; **p.171** *r* © Nathan Benn/Alamy; *b* © Sinclair Stammers/Science Photo Library; **p.172** © digi_dresden – Fotolia; **p.173** Jacques Descloitres, MODIS Rapid Response Team, NASA/GSFC; **p.177** *t* © Geoff Kidd/Science Photo Library; *b* © frans lemmens/Alamy; **p.178** © Dave Bevan/Alamy; **p.180** Enrico Stirl /http://commons.wikimedia.org/wiki/File:Goosenecks_SP4.jpg; **p.181** © Rosemary Mayer/FLPA; **p.183** © Blue Gum Pictures/Alamy; **p.184** © Sinclair Stammers/Science Photo Library; **p.188** © Richard Bizley/Science Photo Library; **p.189** © Photoshot Holdings Ltd/ Alamy; **p.191** © Imagestate Media; **p.192** © Photodisc/Getty Images; **p.193** © Photodisc/Getty Images; **p.202** © Scott Barbour/Getty Images; **p.203** © picturesbyrob/Alamy; **p.204** © Forest Life Picture Library/Isobel Cameron/Crown Copyright; **p.205** © Dr Jeremy Burgess/Science Photo Library; **p.206** © Kazuhiro NOGI/AFP/Getty Images; **p.207** © culture-images GmbH/Alamy; **p.208** *t* © Detlev Van Ravensway/Science Photo Library; *bl* © Horizon International Images Limited/Alamy; *br* Richard Herrmann/OSFPhotolibrary; **p.209** © Philip Dunn/Rex Features; **p.210** © Alex Bartel/Science Photo Library; **p.211** © Robert Harding Picture Library Ltd/Alamy; **p.213** © Andrew Lambert/Science Photo Library; **p.218** *tl* © Photodisc/Getty Images; *tr* © Stockbyte/Getty Images; *c* © Eric Simard – Fotolia; *cl* © Wil Blanche/Rex Features; **p.221** *t* © Iconotec/Alamy; *b* © Jaime Villaseca/The Image Bank/Getty Images; **p.222** © Hartmut SchwarzbachStill pictures/Specialist Stock; **p.225** © James Valentine/Hulton Archive/Getty Images; *b* © Owaki – Kulla/CORBIS; **p.226** *l* © Alt-6/Alamy; *r* © Norbert Schaefer/Alamy; **p.228** *b* © Photodisc/Getty Images; **p.230** Alstom; **p.233** © Stock Connection Distribution/Alamy; **p.236** © Science Photo Library; **p.237** © Digital Vision/Getty Images; **p.239** © Sheila Terry/Science Photo Library; **p.240** © Jean-Loup Charmet/Science Photo Library; **p.242** © Science Photo Library; **p.243** © NASA/Science Photo Library; **p.244** *t* © Matt McPhee – Fotolia; **p.245** © Dr Fred Espenak/Science Photo Library; **p.246** © Space Telescope Science Institute/NASA/Science Photo Library

t = top, *b* = bottom, *l*= left, *r* = right, *c* = centre

Introducing science

Science and scientists

Figure 1 Scientists investigate everything from distant galaxies of stars and the materials here on Earth to the living things around us.

What do you think of when you think of science? Stars and planets? Explosions? Fizzing liquids? Buzzing insects and deep sea creatures? Science is all of these things and more. It is not only a topic but a way of studying. Science is such a huge topic to study that it is divided into three sections:

- **biology** – the study of living things
- **chemistry** – the study of the substances from which things are made (known as **matter**)
- **physics** – the study of how matter and **energy** interact.

Figure 2 Marine zoologists often dive with the animals they are studying. They study in all the oceans of the world.

What kind of scientist would you like to be? Here are just a few examples to think about.

If you went on to study biology you might become a botanist (someone who studies plants) or a marine zoologist (someone who studies sea animals).

Figure 3 This chemical engineer is making an investigation.

If you went on to study chemistry you might become a geochemist (someone who studies rocks to find minerals and oil) or a chemical engineer (someone who makes new substances or designs huge pieces of equipment, called chemical plants, where vast amounts of chemicals are made).

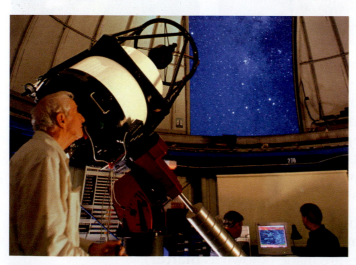

Figure 4 Astronomers use telescopes linked to computers to make their observations.

If you went on to study physics you might become a nuclear physicist finding out more about how matter is made or become an astronomer and study the universe.

For discussion

What kind of scientist would you like to be? Discuss your ideas with your friends and your family.

Scientific equipment

Scientific equipment is called apparatus. There are many kinds of apparatus. Here are just a few examples. You will meet more as you make more and more investigations in this science course.

Apparatus for supplying heat

When scientists want to use heat in an investigation they use a Bunsen burner, or if the laboratory does not have a gas supply they use a spirit burner.

The Bunsen burner

The gas used by many Bunsen burners is methane. As methane burns, it takes part in a chemical reaction with oxygen in the air, and carbon dioxide and water are produced. Both these substances escape into the air as gases.

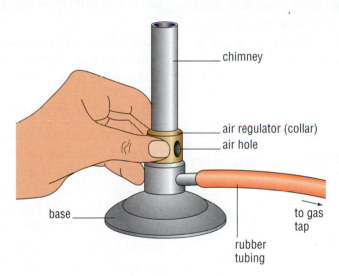

Figure 5 The Bunsen burner

The parts of a Bunsen burner are shown in Figure 5. It is lit and used in the following way.

1 The air regulator or collar must be turned to fully close the air hole before the burner is lit.
2 The match should be lit and placed to one side of the top of the chimney before the gas tap is switched on.
3 When the gas tap is switched on, the gas shoots out through the small hole, called the jet, along the rubber tubing and up the chimney of the burner. Not all the carbon in the gas combines with the oxygen straight away and carbon particles are produced which glow

brightly in the heat and make the flame yellow. If the flame is used to heat anything, the carbon particles form soot on the surface of the apparatus being heated. The flame is called a luminous flame. It is silent. The carbon in the flame reacts with oxygen in the air and forms carbon dioxide.

4 If the collar is now turned and the air hole is fully opened, air mixes with the gas in the chimney. The gases rush up and form a blue cone of unburnt gas at the top of the chimney. Above the cone, methane completely burns away. The flame made when the air hole is completely open is non-luminous and makes a roaring sound.

5 The size of the flame is controlled by the gas tap on the bench. If the tap is fully open a large flame is produced. A smaller flame is produced by partially closing the gas tap.

Less heat energy is released by the luminous flame than the non-luminous flame because the carbon does not all react with oxygen at once. The hottest part of the non-luminous flame is a few millimetres above the tip of the blue cone of unburnt gas.

The spirit burner

In laboratories without a gas supply, spirit burners can be used. The main components of a spirit burner are the reservoir for the fuel and the wick into which the fuel soaks. The fuel is burnt at the top of the wick and is replaced by more fuel from the reservoir until it runs dry. A reservoir may hold enough fuel for the burner to be lit for an hour. Some spirit burners have an adjustable wick to control the rate at which the fuel burns.

General laboratory apparatus

Many pieces of apparatus are made of glass because it is transparent. This makes it easy to see what is happening inside. Glass is also easy to clean. The pieces of apparatus that are to be heated are made from borosilicate glass, also known as Pyrex. This is the same as the glass used for casserole dishes that can be safely put in an oven to cook a meal.

Figure 7 shows the diagrams used to represent some common pieces of apparatus and some photographs of the apparatus in use.

1 Why is one type of flame hotter than the other?

2 Why does closing the gas tap a little reduce the size of the flame?

3 What safety precautions should you take when using a Bunsen burner? Explain the reason for each precaution.

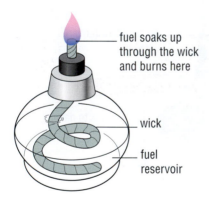

fuel soaks up through the wick and burns here

wick

fuel reservoir

Figure 6 The spirit burner

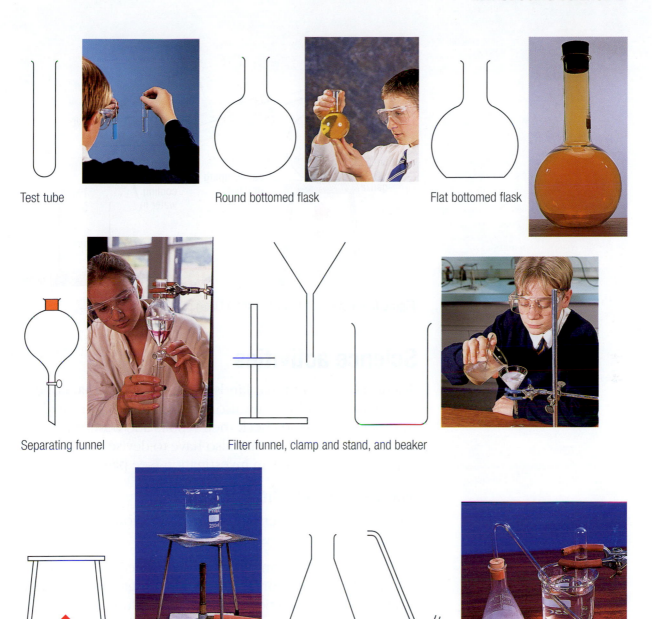

Test tube

Round bottomed flask

Flat bottomed flask

Separating funnel

Filter funnel, clamp and stand, and beaker

Bunsen burner, tripod and gauze

heat

Conical flask with delivery tube

Figure 7 Some common laboratory apparatus

4 Draw the apparatus set up as shown in the bottom left photograph in Figure 7.

5 Draw the apparatus set up as shown in the bottom right photograph in Figure 7.

Some pieces are more complicated. For example, the Liebig condenser is used to condense steam to make water and consists of a tube surrounded by a second tube, called a water jacket, through which cold water flows. Figure 8 shows how pieces of common apparatus and the Liebig condenser are drawn together to make a diagram of how the apparatus is set up. Notice how the stoppers or bungs, shown in orange, are represented with tubes passing through them.

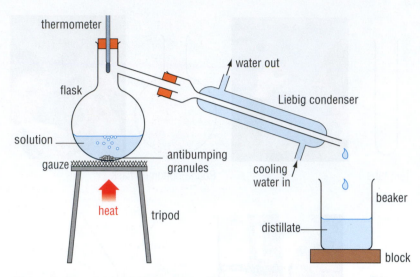

Figure 8 Apparatus for distillation using a Liebig condenser

Science activities

Scientists carry out three kinds of activities. The main one is scientific enquiry but they also repeat investigations made by other scientists to check their results and observations (see page 12) and they may also have to devise new apparatus to carry out an investigation (see page 13).

Making a scientific enquiry

When you make a scientific investigation you are making a scientific enquiry. It has four stages.

- **Stage 1:** Considering ideas and evidence from your own observations or from other investigations. You may also make a prediction at this stage to describe what you think may happen.
- **Stage 2:** Planning the investigation.
- **Stage 3:** Obtaining and presenting evidence. This means making careful observations and measurements and recording them in tables, bar charts and line graphs.
- **Stage 4:** Considering the evidence and the approach. This means drawing conclusions from the investigation and thinking about how the investigation could be improved.

You do not need a lot of apparatus to make a scientific enquiry, as the following example from biology shows.

Stage 1: Considering ideas and evidence

You may notice that many plants release their seeds into the air to spread them away from the parent plant. You may also notice that when a seed falls to the ground it can come to rest in any position – on either side, right way up or upside down. This may lead to you ask the question 'Does the way a seed rests on the ground stop it from sprouting?' You may go further and predict what might happen – for example, you might predict that 'Seeds landing upside down do not sprout'.

Stage 2: Planning

Many wind-blown seeds are very small, and would be difficult to work with to test your idea. So select a seed that is large, with parts that are easy to see, such as the broad bean. This seed has a scar at one end where it was attached inside the pod and a mark on one side from where the root will grow. Plan an experiment in which you take some soaked broad bean seeds and plant them in different positions, as in Figure 9, and then carefully dig them up after a week to see which ones have sprouted.

Figure 9 The beans can be planted in different orientations.

Stage 3: Obtaining and presenting evidence

Record what has happened by making drawings of each bean or by taking a photograph of them.

Stage 4: Considering the evidence and approach

Examine the seeds to see which have sprouted, if any. Then compare your observations with your prediction, decide whether your prediction was correct and state what you have discovered.

Scientific enquiry in more detail

In each of the stages of scientific enquiry, there are activities that can be described as scientific activities. They are shown in the lists that follow.

Ideas and evidence

- Be able to talk about the importance of questions, evidence and explanations.
- Make predictions and review them against evidence.

Figure 10 Discussing ideas for an investigation

Planning

- Suggest ideas that may be tested.
- Outline plans to carry out investigations, considering the **variables** to control, change or observe.
- Make predictions, referring to previous scientific knowledge and understanding.
- Identify appropriate evidence to collect and suitable methods of collection.
- Choose appropriate apparatus and use it correctly.

Figure 11 Planning an investigation

Obtaining and presenting evidence

- Make careful observations, including measurements.
- Present results in the form of tables, bar charts and line graphs.
- Use information from secondary sources.

Figure 12 Collecting evidence and taking measurements

Considering the evidence and approach

- Make conclusions from collected data including those presented in a graph, chart or spreadsheet.
- Recognise results and observations that do not fit into a pattern, including those presented in a graph, chart or spreadsheet.
- Consider explanations for predictions using scientific knowledge and understanding, and communicate these.
- Present conclusions to others using different methods.

Figure 13 Analysing evidence and drawing conclusions

For discussion

Now you have read about scientific enquiry in more detail, how would you set about investigating the question 'Does the way a seed rests on the ground stop it from sprouting?'?

To help you become familiar with these activities, there are Scientific Enquiry spotter questions in every chapter of the book. You can identify them with the icon shown below and by the green background to the question boxes.

When you find one, turn back to pages 7–9 and use the activity lists there to help you answer.

How did scientific enquiry develop?

Making observations

From the earliest times people developed their powers of observation as they hunted animals and gathered nuts, berries and roots. The most successful hunters could not only find tracks from previous observations of their prey but they could tell how old the tracks were and if the animal was fit or injured. Gatherers soon learnt to tell which fruits and vegetables were safe to eat from their colour, smell and taste.

1 How do you think hunters could tell old tracks from new ones?

2 How do you think they could tell if their prey was injured?

Figure A This Iron Age village has been rebuilt by archaeologists. It shows that early people could build strong houses to stand up to the harshness of the winters in northern Europe.

Investigating properties of materials

When the early people wanted to build a shelter as they travelled in search of food, they investigated the properties of the materials around them to discover which would give them protection from the weather. Investigating the properties of materials is a skill used in science that people were using thousands of years ago.

In time many people began to farm the land and settle in one place. They used their knowledge of materials to build permanent homes that would last many years.

3 What sorts of materials would you look for to build a home in a land with cold, wet winters?

Studying energy and forces

Early people also used weapons to hunt. In designing them, they considered two topics of scientific study, although they did not realise it. These topics are energy and forces. For example, they considered the best wood to use to make a bow, which would store plenty of energy as they pulled it back to make the arrow fly. The force generated when the bowstring was released set the arrow on its journey to the target.

Figure B This man from a small tribe of hunter-gatherers living near Lake Eyasi in Tanzania is hunting with a bow and arrow.

4 How could you compare the stored energy in bows made from two different types of wood?

Discussing ideas and observations

Once people had become successful farmers, their settlements grew into towns and cities. The knowledge that had been collected was written down and studied. In Greece, about 2500 years ago, the scholars argued about what was known but most did not believe in testing their ideas with investigations.

Testing ideas – the real start of scientific enquiry

By 1200 years ago, the knowledge built up by the Ancient Greeks and others had passed to the Islamic countries of Africa and Asia, and here the scholars began to test their ideas with investigations – practical scientific enquiry began.

Scientific enquiry starts to spread around the world

By 400 years ago, the practice of scientific investigation and the knowledge it produced had passed to countries in Europe. Scientific enquiry has now moved to almost all parts of the world.

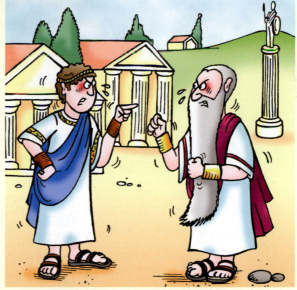

Figure C Greek scholars, called philosophers, built up their own ideas from observations and argued about them. Their students had to decide whose ideas they believed.

5 How long did it take before people began testing the Greeks' ideas with practical scientific enquiry?
6 For how long were Muslim scholars developing practical scientific enquiry before it passed into Europe?

Developing scientific apparatus

As scientists needed to make closer observations, they developed apparatus to help them. One piece of apparatus that has changed greatly over time is the microscope. The first microscopes were little more than wooden tubes with a lens at each end but today they have many interchangeable lenses and are often equipped with lamps and cameras.

Figure D This microscope is fitted with a camera to record the view seen by the scientist.

7 When did people begin making careful observations? Give an example.

8 What do you think might have been appropriate evidence to collect when people were investigating materials to protect them from the weather?

9 Which scientific activities do you think people used to discover which bow sent an arrow the furthest?

Checking another scientist's investigation

Scientists around the world are making investigations all the time. When they complete them, they write them up with their plan and results so that other scientists can try the plan and see if they get the same results. This checking is done to make sure that the information produced by the investigation is true.

Here is an example from chemistry for you to try. A scientist claims to be able to make a hard plastic material from hot milk and vinegar. Here is the plan that was used. Are the scientist's results repeatable?

1 Measure out 300 ml of milk into a beaker and let your teacher warm it for you. The milk should *not* get so hot that it boils.

2 Add 15 ml of vinegar to the milk and stir the two liquids together.

3 Leave the mixture to cool for 15 minutes.

4 Place a kitchen sieve over a second beaker and pour the mixture into the sieve.

5 Squeeze any solid substance in the sieve to remove more liquid.

6 Tip the solid in the sieve onto a paper towel.

7 Examine the substance two hours later, a day later and two days later.

Making equipment for an investigation

Sometimes scientists have an idea for an investigation but do not have the equipment they need to try it. So they have to make the apparatus they need. Try this example from physics in which Galileo, who you will read about later, tested an idea to find out if balls of different masses rolled at different speeds along a slope. He used a water clock to time the journey made by a ball rolling down the slope. A water clock can be made from a bowl with a small hole in the bottom. The water that escapes from the hole is collected in another container. Galileo measured time by weighing the amount of water that had collected in the container while the ball rolled down the slope.

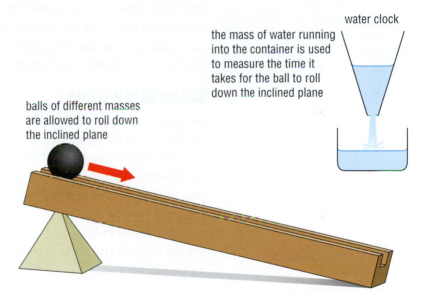

the mass of water running into the container is used to measure the time it takes for the ball to roll down the inclined plane

water clock

balls of different masses are allowed to roll down the inclined plane

Figure 14 The scientific name for the slope Galileo used is an inclined plane.

To check the idea that Galileo tested – that balls of different masses roll at different speeds down a slope – you will need to make a water clock. There are many designs for water clocks. Here are some ideas to try.

- You could make one like Galileo's from an old plastic bowl (ask a teacher to make a hole in it for you) and a container to collect the water. You would also need to work out how you were going to weigh the water using a balance.
- You could put a scale in the collecting container and simply measure the water level on the scale when the investigation is complete.
- You may like to put a float with an upright stick and scale in the bowl and measure how far the float sinks below the rim of the bowl.

For discussion

How could you assess the accuracy of different types of water clock in this investigation?

Try Galileo's experiment

Set up a ramp with a small gradient. Collect two or three balls of the same size but of different masses. Roll each ball down the slope in turn. Time how long each ball takes to make its journey with your water clock. Repeat your experiments to check your results. What do you find?

Safety in the science laboratory

Scientific activities often take place in the laboratory. This is also the room in which apparatus is stored for making scientific investigations. School laboratories are busy places. There may be about 30 people doing investigations in a laboratory at the same time. They may be using gas, water, electricity, a wide range of equipment and some hazardous chemicals.

Laboratory rules

Despite the large amount of activity, there are fewer accidents in laboratories than in most other parts of a school. The reason for this is that when people work in laboratories, they generally take great care to follow the advice of the teacher and the rules pinned to the laboratory wall.

Laboratory rules can be set out in many ways, but should cover the same good advice. On page 15 is an example of a set of rules for working in the laboratory.

Figure 15 It's very important to obey the rules for working in the laboratory, to avoid accidents.

Entering and leaving the laboratory

- Do not run into or out of the laboratory.
- Make sure that school bags are stored safely.
- Put stools under the bench when not in use.
- Leave the bench-top clean and dry.

General behaviour

- Do not run in the laboratory.
- Do not eat or drink in the laboratory.
- Work quietly.

Preparing to do practical work

- Tie back long hair and wear lab coats, if available, buttoned up.
- Wear safety spectacles when anything is to be heated or if any hazardous chemicals are to be used.

During experiments

- Never point a test tube containing chemicals at anyone, and do not examine the contents by looking down the tube.
- Tell your teacher about any breakage or spillage at once. If you are at all unsure of the practical work, check with your teacher that you are following the correct procedure.
- Only carry out investigations approved by your teacher, and use the gas, water and electricity supplies sensibly.

> ## For discussion
>
> **What are the reasons for each of the laboratory rules? What other rules could you add?**

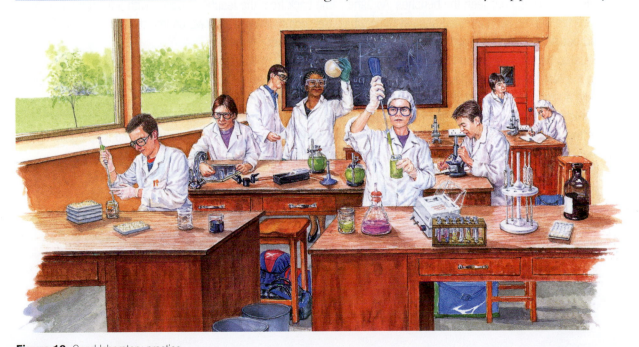

Figure 16 Good laboratory practice

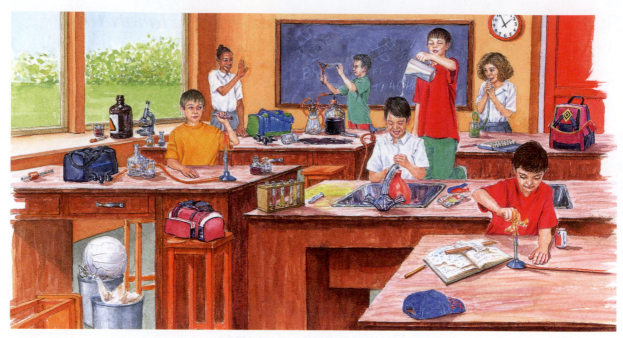

Figure 17 Bad laboratory practice

What are they doing wrong?

Rong ran into the chemistry laboratory because he was keen to do an experiment. He did not see the stool that was sticking out from under the bench and fell over it. He grabbed hold of the bench to stop his fall but his fingers ran into a pool of liquid that had been left on the bench-top and his hand slid, lost its grip and he fell to the floor.

The rest of the class had sat down by the time Rong had picked himself up and put his bag down in the middle of the space between the benches. As Jane came back from the teacher's bench with a lighted taper for her Bunsen burner, she stumbled against Rong's bag. Her long hair swayed forwards into the taper flame. She jerked her head back and only the tips of a few strands of hair were singed.

Eduardo had lit his Bunsen burner and was holding a test tube of liquid over the flame. He was eager to look down the test tube and brushed aside the safety spectacles that Ajani was holding out for him. The liquid boiled quickly; a few drops shot out of the test tube and just missed Eduardo's face.

'Look at that!' he exclaimed, and pointed the test tube at Rong so he could see too. Halima put down the apple that she was secretly eating to see what Eduardo and Rong were doing. When she picked it up again, she did not notice the dark, sticky substance clinging to it that had come from the bench-top.

She quickly put her apple back into her bag, as the teacher approached to check her experiment.

'Did Mrs Jones say to put the apparatus this way round or that way round?' asked Jane, when the teacher had gone away.

'I don't know. I was too busy unsticking my apple from the bottom of my bag,' replied Halima. 'It looks all right like that. Light the Bunsen burner.'

'That's not right!' shouted Akila. Her loud voice made Eduardo jump and he dropped his test tube. Mrs Jones looked round at Akila for a moment, but went off to stop Rong picking up the broken glass with his fingers.

'It should be like ours,' continued Akila in a quieter voice. 'Mrs Jones says it is OK.'

'But Rong's isn't like that,' cried Jane.

'No,' whispered Rong. 'I'm making up my own experiment. If I light this paper in the sink and put this wire behind that dripping tap and just press this switch then ...'

10 List the things that the students are doing wrong in this story.

11 What did the students who were in the laboratory before this class do wrong?

Warning signs

Some of the substances used in scientific investigations are dangerous if not handled properly. The containers of these substances are labelled with a warning symbol such as those shown in Figure 18.

 corrosive

 explosive

 harmful or irritant

 highly flammable

 oxidising

 radioactive

 toxic

6 What do you understand by the each of the following words?
a) corrosive
b) irritant
c) flammable
d) radioactive
e) toxic

Figure 18 Warning symbols

◆ SUMMARY ◆

◆ Science is divided into biology, chemistry and physics (*see page 1*).

◆ Items of scientific apparatus are used in investigations (*see page 3*).

◆ Scientific enquiry is the main activity of scientists (*see page 6*).

◆ Scientists check the results of investigations by repeating them (*see page 12*).

◆ Scientists sometimes have to make equipment for their investigations (*see page 13*).

◆ Rules need to be followed for investigation work in the laboratory to be safe (*see page 14*).

◆ Warning signs are used on the containers of dangerous substances (*see page 17*).

End of chapter questions

1 Arrange the following as a time line of the history of science:
   Greek scholars, worldwide scientists, early people, European scientists, Muslim scientists

2 Which scientists study:
   a) rocks to find minerals and oil
   b) plants
   c) how matter is made
   d) how to make new substances
   e) sea animals
   f) the universe?

3 Arrange the scientists you identified in question **2** into biologists,
   chemists and physicists.

4 A seedling is found that is growing as shown in Figure 19.
   Try your scientific enquiry skills to find out how it came to be growing
   in this way.
   a) Stage 1: If you have tried the bean-growing test in this chapter, what
       information from the results might help you have an idea? If you have
       not done this test, what do you think?
   b) Stage 2: How are you going to test your idea?
   c) Stage 3: How are you going to present your evidence?
   d) Stage 4: What conclusions can you draw from your scientific enquiry?

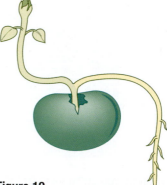

Figure 19

BIOLOGY

1 The characteristics of living things

- ◆ Comparing living things with those that have never lived
- ◆ Signs of life and animals
- ◆ Signs of life and plants
- ◆ Eating and feeding
- ◆ Respiration
- ◆ Movement
- ◆ Irritability
- ◆ Growth and reproduction
- ◆ Excretion
- ◆ Testing for carbon dioxide

Biology is the study of living things. In this chapter, we are going to look at the features or characteristics that something must have for us to identify it as a living thing.

Living and never lived

You can make two groups of things – living things and things that have never lived. The klipspringers in Figure 1.1 are living things, but the rock they are standing on has never lived.

Figure 1.1 Klipspringers live in parts of the African savannah. They spend the hottest part of the day resting among rocks.

1 How is a living thing different from something that has never lived?

For discussion

If you grouped things into living things and things that have never lived, where would you place a block of wood?

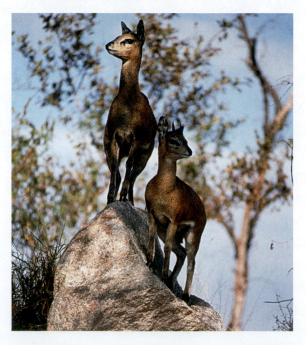

Signs of life

If something is called a living thing, it must have seven special features. These are called the characteristics of life. The characteristics are:

● feeding
● **respiration**
● movement
● growth
● **excretion** (getting rid of waste)
● **reproduction**
● **irritability** (being sensitive to the surroundings).

These activities are also known as life processes.

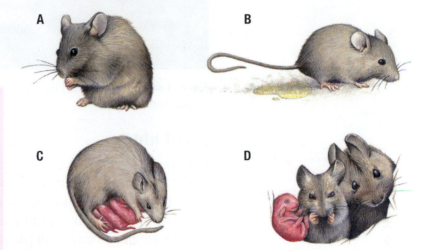

2 Which characteristics of life are shown by the mice in the pictures A–D in Figure 1.2?

3 Does each of the following have any characteristics of life? Explain your answers.
   a) an aeroplane
   b) a computer
   c) a brick

Figure 1.2 Four of the characteristics of life

Figure 1.3 This desert locust is shedding its last skeleton. Here the wings are rolled together, forming an arch on the locust's back.

Animal life

All animals have the same seven characteristics of life but they may show them in different ways. For example, all animals grow, but some have a skeleton on the outside of the body and can grow only when they shed the old skeleton and stretch a new soft skeleton beneath before it sets. Insects and spiders do this by taking in air. Crabs and lobsters stretch their new skeletons by taking in water. Animals with skeletons inside their bodies simply grow larger without having to shed their skeletons.

All living things respire, and most of them use oxygen for this. Many animals living on land have lungs, in which they take oxygen from the air. Many aquatic animals have gills, which take up oxygen dissolved in the water.

Figure 1.4 This axolotl lives in Lake Xochimilco in Mexico. Its gills are on the outside of its body behind its head.

Plant life

Green plants also have the same seven features but they show them in different ways to animals. Plants make food from carbon dioxide in the air and water, by using energy from sunlight. Chemicals in the soil are also needed, but in very small amounts. All plant cells respire and gaseous exchange takes place through their leaves.

4 How is a green plant's way of feeding different from an animal's way of feeding?

For discussion

A car may have five of the characteristics of life. What are they and how does the car show them?

If there are drought conditions, why might a plant produce seeds rather than grow new plantlets?

plantlets

stalk

Figure 1.5 The spider plant grows in many moist woodlands in the warmer regions of the world. It makes plantlets on stalks.

Plants move as they grow and can spread out over the ground. Wastes may also be stored in the leaves. Green plants are sensitive to light and grow towards it. Plants reproduce by making seeds or spores. Some plants can reproduce by making copies of themselves, called plantlets.

Looking at signs of life

Eating and feeding

All living things need food. Plants make their own food but animals must get it from other living things. Some animals, like ourselves, eat a wide range of foods, while others eat only a small range of foods.

In the rainforest ticks, lice, leeches and mosquitoes feed on just one food – blood. They have mouths that can break through skin and suck up their meal. Every animal has a mouth that is specially developed or adapted for the animal to feed in a particular way.

5 How many different kinds of foods do you eat?

6 How is the mouth of a crocodile adapted for feeding?

Figure 1.6 These leeches are being used to draw out blood as part of a medical operation.

Respiration

Respiration is the process in which energy is released from food. The released energy is used for life processes such as growth and movement. Respiration takes place in the bodies of both plants and animals. It is a chemical reaction. During respiration, a food called glucose reacts with oxygen to release energy, and carbon dioxide and water are produced. The word equation for this chemical reaction is:

glucose + oxygen → carbon dioxide + water

Respiration should not be confused with breathing, which is the process of moving air in and out of the body (see page 39). Later, when you study how plants make food, you must remember that while the plants are making food in a process called **photosynthesis** they are also respiring to stay alive.

Movement

Let your right arm hang down by your chair. Stick the fingers of your left hand into the skin in the upper part of your right arm (above the forearm). Raise your right forearm and you should feel the flesh in the upper arm become harder. This is muscle, and it is working to move your forearm upwards. Muscles provide movement for all animals. Animals move to find food, avoid enemies and find shelter. Even when an animal is sitting or standing still, muscles are at work. On page 39 you can see that the diaphragm muscle helps you to breathe, and there are muscles between your ribs that move them up and down. Inside your body your heart muscle pumps blood around the body, and muscles in the wall of your stomach churn up your food to help it digest.

Irritability

Animals detect or sense changes in their surroundings by their sense organs. These are the skin, eyes, ears, tongue and nose. Some animals such as insects and centipedes have long antennae, which they use to touch the ground in front of them. The information their brains receive helps them decide if it is safe to move forwards.

Like many animals, we use our eyes and ears to tell us a great deal about our surroundings. We use our tongue and nose to provide us with information about food. If it smells and tastes pleasant it may be suitable to eat, but if it smells and tastes bad it could contain poisons. The snake shown in Figure 1.7 appears to be tasting the air when it sticks out its tongue, but it is really collecting chemicals in the air, such as scents. It draws its tongue back into its mouth and pushes the tip into a pit in its nose where the chemicals are detected.

Figure 1.7 This grass snake is collecting chemicals in the air with its forked tongue. Grass snakes live in Europe and Northwest Africa.

Growth and reproduction

Living things need food for energy to keep the body alive and for materials. They need the materials for growth and to repair parts of the body that have been damaged. Young animals, like the baby elephants in Figure 1.8, need food to grow healthily.

Figure 1.8 Elephants live in large family groups called herds, ruled by an elderly female called a matriarch. African elephants, like these, have large ears while the Asian or Asiatic elephant has smaller ears.

Once the elephants are fully grown, they need food to keep themselves in good health and to produce offspring. If the elephants did not produce offspring the herd would eventually disappear as the old elephants died. **Reproduction** is the process that keeps a plant or animal species in existence.

Excretion

When food and oxygen are used up in the body, waste products are made. These are poisonous, and if they build up inside the body they can kill it. To prevent this from happening, the body has a way of getting rid of its harmful wastes. It is called **excretion**. Wastes are released in urine, sweat and the air that we breathe out. The waste product we release in our breath is carbon dioxide.

Testing for carbon dioxide in exhaled breath

You can test your exhaled air for carbon dioxide by passing it through limewater (Figure 1.9). If carbon dioxide is present it reacts with the calcium hydroxide dissolved in the water to produce insoluble calcium carbonate. This makes the water turn white or milky.

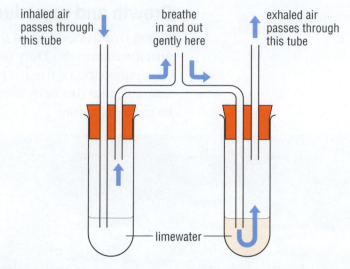

Figure 1.9 Testing inhaled and exhaled air for carbon dioxide

Testing for carbon dioxide in air around seeds

Carbon dioxide production can be used as an indication of respiration and a sign of life. Hydrogen carbonate indicator is a liquid that changes colour in the presence of carbon dioxide. It changes from an orange-red colour to yellow. The production of carbon dioxide by germinating pea seeds can be shown by setting up the apparatus in Figure 1.10.

Figure 1.10 Investigating carbon dioxide production by germinating pea seeds

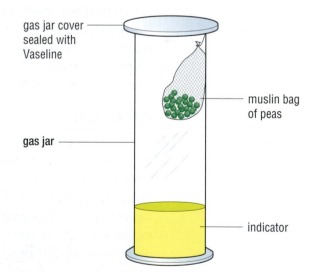

Looking for life beyond the Earth

Almost everyone has an opinion about alien life in space but how do scientists go about investigating it? In the 1970s, the Viking missions to Mars took place and this provided scientists with a chance to devise a scientific investigation to test for signs of life.

In the first stage of the investigation, scientists thought about the link between food and respiration. They reasoned that if living things were present in the soil they might be detected in the following way. Water containing food could be put into the soil. If living things were present in the soil, they would feed on the food and produce carbon dioxide in respiration. So if carbon dioxide was detected in the soil, living things were present. Apparatus was then designed to dig up the soil and put it in a container where food and water was added. As Mars is much colder than Earth, a heater was built around the container to warm up the soil and make the living things, if they were present, feed and respire faster.

Figure A A Viking lander on Mars with its digging tool ready to scoop up Martian soil for testing

When the spacecraft reached Mars, the soil was tested and carbon dioxide gas was detected. It seemed that the gas had been produced by living things, but when the evidence from other investigations taking place there was considered, the scientists concluded that there could be other explanations for the gas being produced. More missions to Mars are planned in the next few years to re-test the soils for living things.

Scientists are also planning to look for living things on a moon of Jupiter called Europa. In the first stage of their investigation, they are considering the following evidence to support their idea. **Space probes** indicate that there is a strong possibility of oceans of water on Europa, beneath a thick icy surface. We know that water is needed for life on Earth. Deep in the Earth's oceans, hot springs of black water have been found, called black smokers. The water contains large amounts of a substance called hydrogen sulphide. Some tiny forms of living things have been found which use this substance for food. Perhaps there are black smokers and tiny living things on Europa.

Figure B A black smoker on the ocean floor

1 The scientists used feeding and respiration as characteristics of life to investigate. How could irritability and movement be used to plan an investigation?

2 Did the investigations show that the gas definitely did *not* come from living things? Explain your answer.

3 What kind of equipment would need to be carried on a space probe to Europa to look for life on its ocean floor?

4 In planning the first stage of the investigation, what was the idea to be tested?

5 What pieces of evidence for life would help you decide how to plan an investigation to look for life on Europa?

◆ **SUMMARY** ◆

◆ There are seven characteristics of life – feeding, respiration, movement, growth, excretion, reproduction and irritability (*see page 21*).
◆ Green plants make food from carbon dioxide and water (*see page 22*).
◆ Animals must obtain food from other living things (*see page 23*).
◆ Energy for life processes is released in respiration (*see page 23*).
◆ Muscles provide movement for all animals (*see page 24*).
◆ Sense organs are used to detect changes in the environment (*see page 24*).
◆ Food is needed for growth (*see page 25*).
◆ Plant and animal species stay in existence through reproduction (*see page 25*).
◆ Excretion is the release of harmful waste products (*see page 25*).
◆ Limewater is used to test for carbon dioxide (*see page 25*).
◆ Hydrogen carbonate indicator is used to test for carbon dioxide (*see page 26*).

End of chapter questions

The apparatus shown in Figure 1.11 is set up to show that seeds use up oxygen when they respire. The soda lime absorbs any carbon dioxide in the tube.

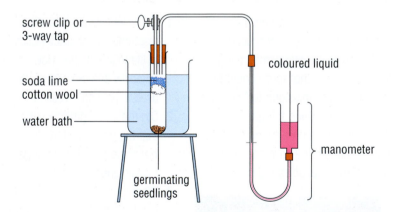

Figure 1.11

1 What happens to the coloured liquid in the tube as the seedlings respire?

2 Why does the volume of the gas around the seedlings change?

 3 How would you use boiled seedlings to show that any change of gas was due to respiration?

◆ Organisms and organs
◆ The organs of flowering plants
◆ The organ systems of the human body
◆ The skeletal system
◆ The muscle system
◆ The circulatory system
◆ The respiratory system
◆ The digestive system
◆ The nervous system
◆ The excretory system
◆ The sensory system
◆ The endocrine system
◆ Scientists and the human body
◆ Exercise

Living things are also called living **organisms**. They are given this name because most living things have bodies composed of organs. An **organ** is a part of the body that performs a special task to help the organism live. The task an organ performs is related to one or more of the seven life processes (see page 21). Some organs form a group to carry out a task to keep a living thing alive. These groups of organs are called **organ systems**.

In flowering plants, the major organs are easy to see because they form separate parts that make up the external appearance of the organism. In animals, most of the major organs of the body are enclosed inside it and cannot be seen.

Organs of a flowering plant

There are five main organs in the body of a flowering plant. They are the root, stem, leaf, flower and bud. Each organ may be used for more than one task or life process.

1 The **root** anchors the plant and takes up water and minerals from the soil. The roots of some plants, such as the carrot, store food.

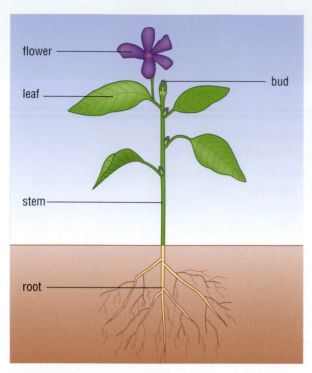

Figure 2.1 Organs of a flowering plant

2 The **stem** transports water and food and supports the leaves and the flowers. Some plants, such as trees, store food in their stems.

3 The **leaf** produces food. In some plants, such as the onion, food is stored in the bases of the leaves. The swollen leaf bases make a bulb.

4 The **flower** contains the reproductive organs of the plant.

5 The **bud** contains tiny new branches, leaves and flowers ready to grow. These are delicate structures so the outside of the bud forms a protective covering to prevent them being damaged as they start to grow.

All the organs work together to keep the plant alive so that it can grow and produce offspring.

1 Draw a table featuring the organs of a flowering plant and the tasks they perform.

2 How is the leaf dependent on the root and the stem?

3 Which life processes or tasks do you think are found in both plants and humans? Explain your answer.

A closer look at plant organs

The root

The part of the plant we call the root is really made from a number of separate roots. They form a root system. There are two main types.

● In the tap root system, a main root – called a tap root – grows downwards, while lateral roots grow downwards and sideways.

● In the fibrous root system, all the roots are the same size and can grow downwards and sideways.

Roots can develop with special features to allow a plant to survive in a certain place. For example, orchids in rainforests grow on the branches of the trees and their roots hang down in the air to collect water from raindrops or mist.

Figure 2.2 The fibrous root system of a grass plant, and the tap root system of a dandelion plant

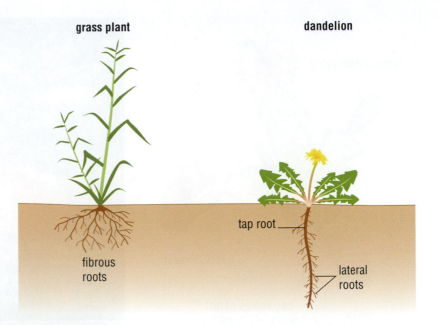

grass plant

dandelion

tap root

fibrous roots

lateral roots

Figure 2.3 The pneumatophores of these mangrove trees, growing in Indonesia, help their roots in the mud get the oxygen they need.

All roots need oxygen and they normally get it from the air spaces in the soil. In swamp mud there is little oxygen, so the roots of the mangrove trees that live in such habitats have special structures called **pneumatophores**, which grow upwards to the surface and take in oxygen from the air for the roots to use.

Figure 2.4 The tendril of this pea plant is wrapping around the stem of another plant.

The stem

The stems of most plants grow upwards and support themselves. Some plants have stems that are too weak to support themselves, and instead grow up the sides of other, larger plants. They may have structures called **tendrils**, which look like springs, growing and curling around the parts of the supporting plant.

A few plants have stems that grow across the ground or even under it. The ginger plant has an underground stem in which it stores food.

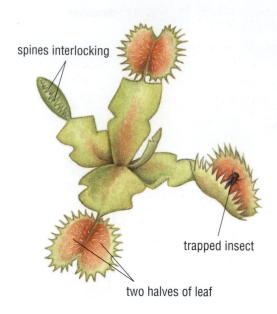

spines interlocking

trapped insect

two halves of leaf

Figure 2.5 The Venus fly trap lives in swamps in parts of North America.

Figure 2.6 These plants growing in a field in Europe have buds of different shapes and flowers of different colours.

For discussion

How can the stems, leaves, flowers and buds of plants be used to identify them? Look at a few plants that are grown as house plants to find out.

4 Which scientific enquiry skills do you use in the investigation suggested in the box above?

Leaves

Leaves can have many shapes. They can be long and thin, spear shaped, rounded, oval or divided into lobes. Some, such as holly leaves, have spines. A very few plants have leaves that can catch insects – they close quickly over the insect when it settles on them.

Flowers

A huge number of plants use insects to help them reproduce and have developed brightly coloured flowers to attract them. The flowers can be shaped like a bell, a star, a cross, a saucer, a tube or even a trumpet.

Buds

Plant buds vary in size, shape and colour and some buds, particularly tree buds, can be used to identify a plant.

Organ systems of a human

In animals, there are groups of organs that work together to carry out a task to keep the animal alive. These groups are called **organ systems** and each system performs tasks related to one or more of the life processes (see page 21):

- skeletal system
- muscle system
- circulatory system
- respiratory system
- digestive system
- nervous system
- excretory system
- sensory system
- endocrine system.

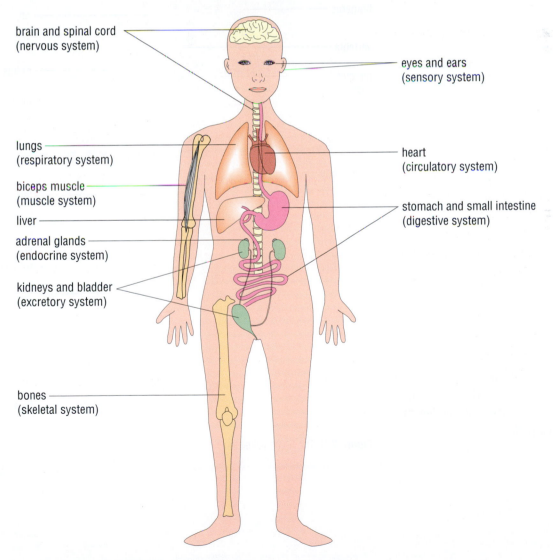

brain and spinal cord
(nervous system)

eyes and ears
(sensory system)

lungs
(respiratory system)

heart
(circulatory system)

biceps muscle
(muscle system)

liver

adrenal glands
(endocrine system)

stomach and small intestine
(digestive system)

kidneys and bladder
(excretory system)

bones
(skeletal system)

Figure 2.7 Organs in the human body

A closer look at the skeleton and muscles

There are 206 bones in the human skeleton. Each arm and hand together have 30 bones. Each leg and foot together have 30 bones. The skeleton accounts for 15% of the mass of the body. The tissue of the skeleton (bone) is hardened as it takes calcium from the digested food.

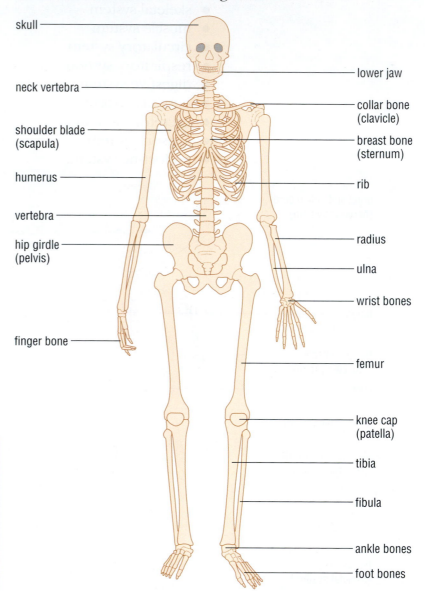

skull

neck vertebra

shoulder blade (scapula)

humerus

vertebra

hip girdle (pelvis)

finger bone

lower jaw

collar bone (clavicle)

breast bone (sternum)

rib

radius

ulna

wrist bones

femur

knee cap (patella)

tibia

fibula

ankle bones

foot bones

Figure 2.8 The human skeleton

5 There are three bones in the arm. How many are in the wrist and hand?

6 How many bones are there in all four limbs?

7 Figure 2.8 shows the main bones of the body. How many of these bones can you feel in your body?

The skeleton and protection

The brain and the spinal cord form the central nervous system and are made from soft tissue. They could be easily damaged without a hard covering. The bones of the skull are fused together to make a strong case around the brain. The backbone is made of 33 bones known at vertebrae (*singular*: vertebra). There is a hole in each vertebra through which the spinal cord runs. The column of vertebrae makes a tube of bone around the spinal cord. There are gaps between the vertebrae through which nerves pass from the spinal cord to the body. The ribs and backbone form a protective structure around the lungs and heart.

8 The skull forms a solid sheet of protection and the ribs form a cage. Why do you think the rib cage is not a solid sheet like the skull? Which offers better protection, the sheet or the cage? Explain your answer.

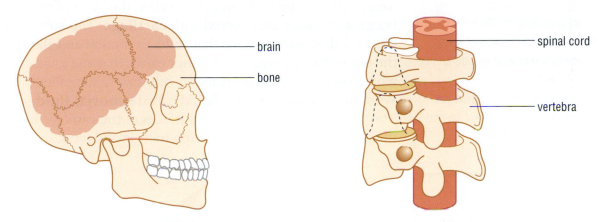

Figure 2.9 Protection of the central nervous system: the brain (left) and the spinal cord (right)

The skeleton and support

The organs that form systems such as the digestive, circulatory, excretory and respiratory systems account for 20% of the body's weight. The organs are made from soft material and have no strong supporting material inside them. The bones of the skeleton provide a strong structure to which the organs are attached. They allow the organs to be spread out in the body without squashing into each other. The muscles account for 45% of the body's weight. They are also made from soft tissue but gain their support from the bones to which they are attached.

9 A person has a mass of 43.5 kg. How much of this mass is due to:
a) their organ systems
b) their muscles?
10 The percentage of the body's mass not accounted for by the skeleton, organs and muscles is due to fat. What percentage of the body's mass is due to fat?

The skeleton and movement

The place where bones meet is called a **joint**. In some joints, such as those in the skull, the bones are fused together and cannot move. Most joints, however, allow some movement. Some joints, such as the elbow or knee, are called hinge joints because the movement is like the hinge on a door. The bones can only move forwards or backwards. A few joints, such as the hip, are called ball-and-socket joints because the end of one bone forms a round structure like a ball that fits into a cup-shaped socket. These joints allow much more movement.

To stop the bones coming apart when they move, they are held together by fibres called **ligaments**. To stop them wearing out as they rub over each other, the parts of the bones in the joint are covered with **cartilage**. This substance has a hard, slippery surface that reduces friction and allows the bones to move over each other easily. In some joints, where there is a lot of movement, cells in a tissue called synovial membrane make a liquid called **synovial fluid**. This fluid spreads out over the surfaces of the cartilage in the joint and acts like oil, reducing friction and wear.

11 Look at the skeleton in Figure 2.8 (page 34). Which bones meet at:
  a) the hip joint
  b) the knee joint
  c) the elbow joint
  d) the shoulder joint?
12 Name:
  a) two hinge joints
  b) two ball-and-socket joints.
13 How might a joint be affected by:
  a) torn ligaments
  b) lack of synovial fluid
  c) damaged cartilage?
14 Why do you think that some joints are painful in elderly people?
15 How does the body stop you using a damaged joint so that it has time to heal?

For discussion

What would the body be like without a skeleton?

Could the body survive without a skeleton?

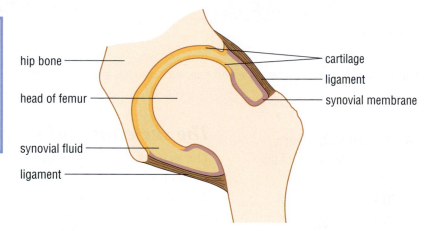

hip bone

head of femur

synovial fluid

ligament

cartilage

ligament

synovial membrane

Figure 2.10 Inside a hip joint

Muscles

Muscle is made up from tissue that has the power to move. It can contract to become shorter. A muscle is attached to two bones across a joint. When muscle gets shorter, it exerts a pulling force. This moves one of the bones but the other stays stationary. For example, the biceps muscle in the upper arm is attached to the shoulder blade and to the radius bone in the forearm. When the biceps shortens or contracts it exerts a pulling force on the radius and raises the forearm.

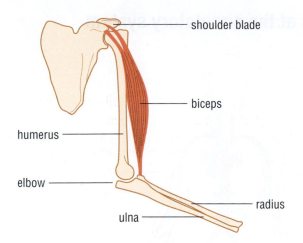

Figure 2.11 Biceps on arm bones

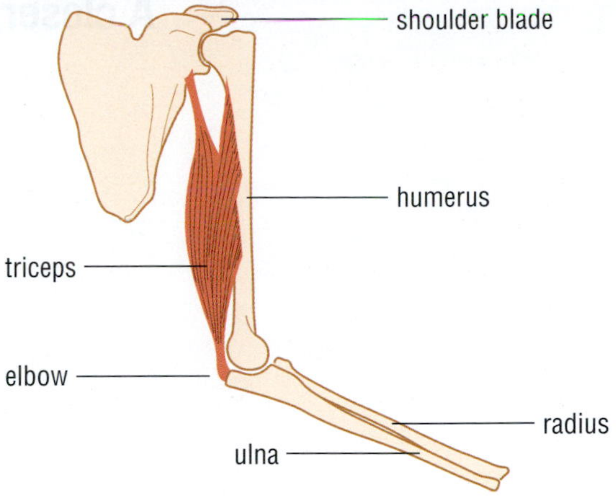

Figure 2.12 Triceps on arm bones

16 Draw a diagram featuring both the biceps and the triceps, showing the triceps fully shortened.

17 Using dotted lines, draw on the position of the forearm when the biceps is fully shortened.

Figure 2.13 The muscle raising this weight is at the front of the upper arm and is called the biceps.

18 What movements take place in the body that you do not have to think about?

A muscle cannot lengthen or extend itself. It needs a pulling force to stretch it again. This force is provided by another muscle. The two muscles are arranged so that when one contracts it pulls on the other muscle, which relaxes and lengthens. For example, in the upper arm the triceps muscle is attached to the shoulder blade, humerus and ulna. When it contracts, the biceps relaxes and the force exerted by the triceps lengthens the biceps and pulls the forearm down. When the biceps contracts again, the triceps relaxes and the force exerted by the biceps lengthens the triceps again and raises the forearm. The action of one muscle produces an opposite effect to the other muscle and causes movement in the opposite direction. The two muscles are therefore called an **antagonistic muscle pair**.

There are two other kinds of muscle – smooth muscle and cardiac muscle. Smooth muscle is found in other organ systems, such as the digestive system, where it moves the food along the alimentary canal. The heart is made of cardiac muscle. Its movement pumps the blood around the body. The nervous system controls the movement of these two kinds of muscle automatically so you do not have to think about them.

A closer look at the circulatory system

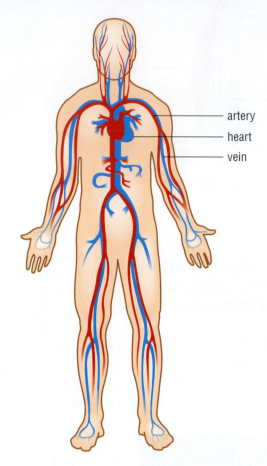

Figure 2.14 The heart and major arteries and veins

The heart is located near the centre of the chest. It is made of muscle called cardiac muscle, which makes the heart beat. As the heart beats it pushes blood into the arteries and draws blood in from the veins. The beating of the heart makes the blood circulate around the body. The heart and blood vessels make up the **circulatory system**.

The beating of the heart can be checked by taking the pulse. You can find your pulse by following these instructions.

1 Hold out your right hand with the palm up.
2 Put the thumb of your left hand under your wrist.
3 Let the first two fingers of your left hand rest on the top of your wrist.
4 Feel around on your wrist with these two fingers to find a throbbing artery. This is your pulse.
5 You can measure your pulse rate by counting how many times your pulse beats in a minute.

Figure 2.15 Measuring a pulse

A closer look at the respiratory system

Breathing

Put your hands just below your ribs as you are reading these next few lines (as in Figure 2.16). As you read, you should feel your hands moving in and out. They are being pushed by a muscle called the **diaphragm**. This helps you to breathe. If you stand up and place your hands on your ribs while breathing in and out you should feel your ribs move. They also help you to breathe. The number of breaths you take in a certain time is called your rate of breathing. You take one breath when you breathe in and out once. The oxygen taken in during breathing is used to release energy from food.

If you have tried the breathing activity above you have already felt the action of part of your **respiratory system** – the ribs and diaphragm. They are the parts that ventilate the system – that is, they move air in and out of it. Air enters through the nose, passes down the back of the mouth and into the voice box and windpipe. You can feel your voice box and windpipe by placing your fingers on the front of your neck. They are hard because they are made of cartilage, or gristle. This material helps to keep the airways open at all times.

The bottom of the windpipe divides into two tubes called bronchi (*singular*: **bronchus**). The bronchi carry the air into the lungs. Here, some oxygen passes through the walls of the lungs into the blood. Carbon dioxide passes from the blood through the walls of the lungs into the air.

Figure 2.16 Feeling the movements made in breathing

19 How does your pattern of breathing change when you exercise and then rest? Try a simple investigation to find the answer.

20 Which scientific enquiry skills do you use in an investigation to see how your pattern of breathing changes when you exercise and then rest?

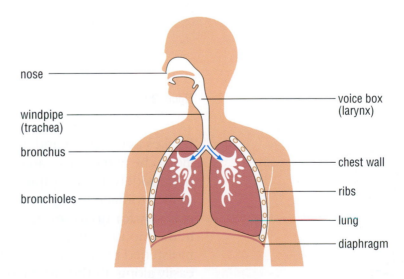

Figure 2.17 The respiratory system

A closer look at the digestive system

The main part of the **digestive system** is a tube that runs through the body. It is called the **alimentary canal**, and includes the oesophagus, stomach and intestines (Figure 2.18). In an adult it is about 8–9 m long. If you wind out a thread from a ball of wool until it is 9 m long, you will get an idea of the length of the alimentary canal. If you fold up the thread you will see that it can fit into a small space. The folding of the alimentary canal allows it to fit in the lower part of the body called the abdomen.

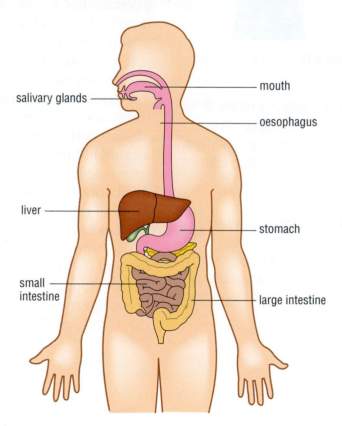

salivary glands

mouth

oesophagus

liver

stomach

small intestine

large intestine

Figure 2.18 The digestive system

It takes between 24 and 48 hours for the food to travel along the alimentary canal, but journey times can vary quite widely. For example, a meal of boiled rice only stays in the stomach for up to two hours, while roast chicken may stay for up to seven hours. Food begins its journey in the mouth, where it is broken up by the teeth and moistened by the saliva so that the small pieces can slide easily along. In the stomach, the food is churned up into a creamy liquid called **chyme**, before it continues its journey into the intestines.

A closer look at the nervous system

The **nervous system** comprises the brain, spinal cord and nerves. The brain is enclosed in the skull and the spinal cord is enclosed in the spinal column (backbone). Nerves connect the brain to the eyes, ears, tongue, nose and skin on the head. Nerves also connect the spinal cord to the skin and to other organs in the body.

Messages travel through the nervous system as tiny electrical signals. The sense organs send signals to the spinal cord and the brain. If the brain decides that the body should move, it sends signals to the muscles.

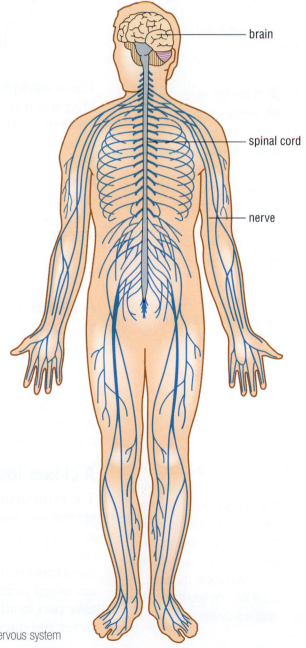

— brain

— spinal cord

— nerve

Figure 2.19 The nervous system

A closer look at the excretory system

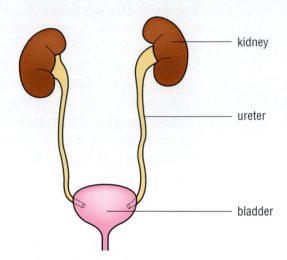
kidney

ureter

bladder

Figure 2.20 Organs involved in the excretion of urine. The urine from the kidneys is stored in the bladder before it is excreted.

21 What are the sense organs in the sensory system?
22 Which organ system:
   a) transports materials around the body
   b) absorbs food into the blood
   c) detects changes in the environment
   d) coordinates activities
   e) takes in oxygen from the air
   f) supports the body
   g) removes waste from the blood
   h) moves bones?
23 Which organ system or systems are involved in:
   a) movement
   b) nutrition
   c) circulation?
24 You are walking across a road and hear a sound behind you. You turn and see that a car has swerved to avoid a donkey and is heading straight for you. What body systems work to get you out of the car's way? Why do you think these systems developed?

The wastes produced by the body collect in the blood. They are removed from the body by the **excretory system**. As the blood passes through the kidneys, a waste called **urea** is filtered from the blood with some water. This mixture of urea and water is called **urine**. As the blood passes through the skin on a hot day, water and a little urea are taken from it and released onto the skin. The main purpose of this action is to cool the skin, although a little urea is also excreted. As the blood passes through the lungs, carbon dioxide is removed and passed into the air, ready to be breathed out.

A closer look at the sensory system

The **sensory system** is made up of the sense organs – the eyes, ears, nose, tongue and skin are the sense organs of sight, hearing, smell, taste and touch respectively. The function of this system is to provide information about the surroundings of an animal. The information is sent in the form of electrical signals or messages along nerves in the nervous system to the brain.

A closer look at the endocrine system

The **endocrine system** is made up of glands, which release chemicals called **hormones** into the blood. The adrenal gland is an example of an endocrine gland. It is found just above the kidney and releases (or secretes) a hormone called adrenaline. You may feel the effect of adrenaline if you are asked to read aloud or act in front of a large audience, or take part in athletics, for example. It makes your heart beat faster and directs more blood to your muscles.

25 What is the name of the hormone that makes your heart beat faster and directs more blood to your muscles?

Hormones also control the way people grow and develop. The hormone insulin helps the body store a sugar that has been absorbed from digested food. A lack of this hormone in the body leads to a disease called **diabetes**. Diabetes can be controlled by taking extra insulin into the body.

Scientists and the study of the human body

The earliest studies on the human body must have occurred back in the Stone Age when people looked for cures for their various ailments and injuries, although we have no records of this. We know for certain that people were developing a knowledge of the human body and medicine by the time of the Ancient Egyptians because of documents written on papyrus that have been discovered. The documents show that the Egyptians knew about the heart, liver, kidneys and bladder and how to treat injuries to the head and upper body.

Galen, who lived 1800 years ago, was a scholar in Ancient Greece. He gathered the writings of previous work on the human body and made his own discoveries. He was not allowed to dissect human bodies so he dissected animals such as monkeys. His work was used by scholars for over a thousand years.

Ibn al-Nafis, who lived 800 years ago in Egypt, studied many subjects. Among them was the study of the human body. He made dissections of the human body and discovered how it differed from the bodies Galen had observed. The books he wrote to explain his work had diagrams in them to make the arrangement of the body easier to understand.

Andreas Vesalius (1514–1564) was a Flemish doctor who dissected human bodies and had an artist make drawings of his work. After he made his discoveries and set out his ideas, he checked them again to make sure of his work. He encouraged his students to do the same – repeat the work to make sure the conclusions are correct. The work and ideas of Vesalius helped scientists make further discoveries in the following centuries.

1 Why is it that we can't be sure about what Stone Age people knew but we can be more certain about the Ancient Egyptians' knowledge?
2 Why do you think Galen's work about the human body was sometimes inaccurate?
3 What did Ibn al-Nafis use that made his work easier to understand?
4 How did Vesalius make his investigations more accurate?

5 Which scientific activity do you think is shown by the production of documents on papyrus in Egypt?
6 Did Galen use information from secondary sources? Explain your answer.
7 What secondary sources do you think Ibn al-Nafis used and why did he use them?
8 Which scientific activities did Vesalius's artist use?

Figure A A drawing of Versalius's work to show the arrangement of muscles in the human body

Today scientists are still investigating the organs of the human body in **research programmes** around the world. These take place in universities and hospitals where the work involves not just looking at the structure of the organs, as the earlier scientists did, but also looking at how they work, how they are damaged by disease and how they can be repaired.
The results of their research can be used by doctors and nurses to improve the chances of sick people returning to full health.

9 Why could using information from secondary sources be a good starting point to set up a research programme?

For discussion

Look back through this chapter at the different organ systems. If you were a research scientist which organ system would you want to study and why would you want to study it?

◆ SUMMARY ◆

◆ The five main organs of the flowering plant are the root, stem, leaf, flower and bud (*see page 29*).

◆ The major organ systems of the human body are:
 – the skeletal system (bones and joints) (*see page 34*)
 – the muscle system (muscles and joints allow bones to move) (*see page 36*)
 – the circulatory system (*see page 38*)
 – the respiratory system (*see page 39*)
 – the digestive system (*see page 40*)
 – the nervous system (*see page 41*)
 – the excretory system (*see page 42*)
 – the sensory system (*see page 42*)
 – the endocrine system (*see page 42*)

End of chapter questions

1 Name three organs of a flowering plant that may be used to store food.

2 What action would you take to check your heart beat?

3 Name four structures in your body that air passes through as it moves to the lungs.

4 In which organ system of the human body is:
 a) the adrenal gland
 b) the bladder
 c) the liver?

◆ How cells were discovered
◆ The microscope
◆ The basic parts of a cell
◆ Adaptations in cells

In Chapter 2, we saw how the body is divided into systems of organs. In this chapter, we are going to find out what organs are made from.

From organs to cells

In the 16th century, scientists in Europe began studying the human body by dissection. This work has continued ever since.

Figure A This painting by Rembrandt shows Dr Nicolaes Tulp making a dissection, in Holland in 1632.

Marie-François X. Bichat (1771–1802) was a French doctor who did many post mortem examinations. In the last year of his life, he carried out 600. He cut up the bodies of dead people to find out how they had died. From this he discovered that organs were made of layers of materials. He called these layers 'tissues' and identified 21 different kinds. For a while, scientists thought that tissues were made of simple non-living materials.

In 1665, long before Bichat was born, an English scientist named Robert Hooke (1635–1703) used a microscope to investigate the structure of a very thin sheet of cork. He discovered that it had tiny compartments in it. He thought of them as rooms and called them 'cells', after the small rooms in monasteries where monks worked and meditated.

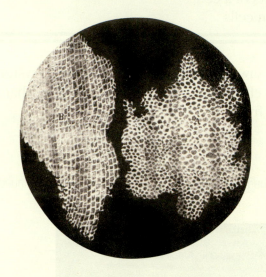

Figure B Here are the compartments in cork that Hooke saw using his microscope. He called them cells.

Bichat did not examine the tissues he had found under a microscope because most of those made at that time did not produce very clear images. When better microscopes were made, scientists investigated pieces of plants and found that, like cork, they also had a cell structure. The cells in Hooke's piece of cork had been empty but other plant cells were found to contain structures.

A Scottish scientist called Robert Brown (1773–1858) studied plant cells and noticed that each one had a dark spot inside it. In 1831, he named the spot the nucleus, which means 'little nut'.

Matthias Schleiden (1804–1881) was a German scientist who studied the parts of many plants and in 1838 he put forward a theory that all plants were made of cells. A year later Theodor Schwann (1810–1882), another German scientist, stated that animals were also made of cells.

The ideas of Schleiden and Schwann became known as the Cell Theory. It led other scientists to make more discoveries about cells and to show that tissues are made up of groups of similar cells.

1 Where did Bichat get his ideas that organs were made from tissues?
2 Who first described 'cells' and where did the idea for the word come from?
3 Who named the nucleus and what does the word mean?
4 What instrument was essential for the study of cells?
5 How could the Cell Theory have been developed sooner?
6 Arrange these parts of a body in order of size starting with the largest: cell, organ, tissue, organ system.

7 Which scientific activity did Bichat perform to think up his idea that organs were made from tissues?
8 Which was the final scientific activity that led Schleiden and Schwann to set up their Cell Theory?

The microscope

Most laboratory microscopes give a magnification up to about 200 times but some can give a magnification of over 1000 times. The microscope must also provide a clear view, and this is achieved by controlling the amount of light shining onto the specimen.

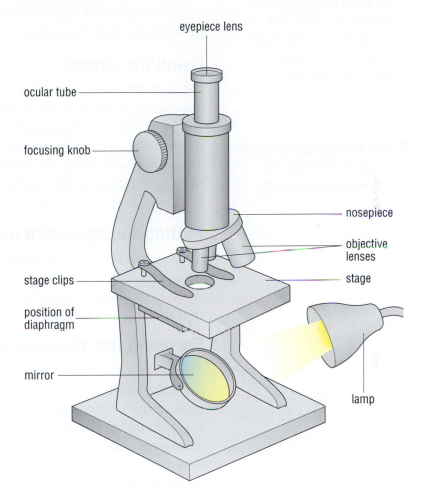

Figure 3.1 The main parts of a microscope

The microscope and light

Light is collected by a mirror at the base of the microscope. The mirror is held in special joints that allow it to move in any direction. The light comes from a lamp or from a sunless sky. It must never be collected directly from the Sun as this can cause severe eye damage and blindness. Some microscopes have a built-in lamp instead of a mirror. The light shines directly through a hole in the stage onto the **specimen**.

1 What is a microscope used for?
2 What advice would you give someone about how to collect light to shine into a microscope?

3 What magnification would you get by using an eyepiece of ×5 magnification with an objective lens of ×10 magnification?

4 If you had a microscope with ×5 and ×10 eyepieces and objective lenses of ×10, ×15 and ×20, what powers of magnification could your microscope provide?

5 How would you advise someone to use the three objective lenses on the nosepiece?

6 Why should you not look down the microscope all the time as you try to focus the specimen?

7 Look at the picture of the microscope on page 47 and describe the path taken by light from the lamp near the microscope to the user's eye.

8 If there are about 6000 million people on the Earth, how many cells have you got in your body?

The magnification of the microscope

Above the stage is the ocular tube. This has an eyepiece lens at the top and one or more objective lenses at the bottom. The **magnification** of the two lenses is written on them. An eyepiece lens may give a magnification of ×5 or ×10. An objective lens may give a magnification of ×10, ×15 or ×20. The magnification provided by both the eyepiece lens and the objective lens is found by multiplying their magnifying powers together.

Using the lenses

Most microscopes have three objective lenses on a nosepiece at the bottom of the ocular tube. The nosepiece can be rotated to bring each objective lens under the ocular tube in turn. An investigation with the microscope always starts by using the lowest-power objective lens, and then working up to the highest-power objective lens if it is required.

Putting the specimen under the microscope

A specimen for viewing under the microscope must be put on a glass slide. The slide is put on the stage and held in place by the stage clips. The slide should be positioned so that the specimen is in the centre of the hole in the stage.

Focusing the microscope

The view of the specimen is brought into focus by turning the focusing knob on the side of the microscope. This may raise or lower the ocular tube, or it may raise or lower the stage. In either case you should watch from the side of the microscope as you turn the knob to bring the objective lens and specimen close together. When the objective lens and the specimen are close together, but *not* touching, look down the eyepiece and turn the focusing knob so that the objective lens and specimen move apart. If you do this slowly, the blurred image will become clear.

Looking at cells

There are ten times more **cells** in your body than there are people on the Earth. If you stay in the water in a swimming pool for a long time you may notice sometimes that when you dry yourself part of your skin flakes off. These flakes are made of dead skin cells. You are losing skin cells all the time but in a much smaller way.

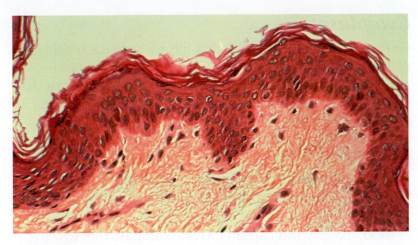

Figure 3.2 Section of human skin. Cells can be seen flaking off the surface.

As your clothes rub against your skin they pull off tiny flakes, which pass into the air and settle in the dust. A small part of the dirt that cleaners sweep up at the end of a school day comes from the skin that the students have left behind.

Figure 3.2 shows a section of human skin that has been stained and photographed down a microscope using a high-power objective lens. When unstained, the different parts of the cells are colourless and are difficult to distinguish. In the 1870s, it was discovered that dyes could be made from coal tar which would stain different parts of the cell. Cell biologists found they could stain the nucleus and other parts of the cell different colours to see them more easily.

9 Why are most specimens of cells stained before they are examined under the microscope?

Basic parts of a cell

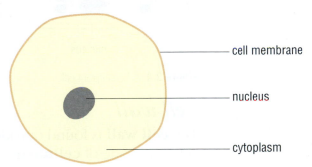

cell membrane

nucleus

cytoplasm

Figure 3.3 A typical animal cell

Nucleus

The **nucleus** is the control centre of the cell. It contains the **genetic material**, called **DNA** (its full name is deoxyribonucleic acid). DNA is made from groups of **atoms** linked together to form a **molecule** like a long chain. The groups of atoms occur in different combinations along the DNA molecule. The combinations of the groups provide instructions for the cell to make chemicals to keep it alive or to build its cell parts. As a cell grows the DNA is copied, and when the cell divides the DNA divides too, so that the nucleus of each new cell receives all the instructions to keep the new cell alive and enable it to grow.

10 Imagine that you are looking down a microscope at a slide labelled 'Cells'. You can see a coloured substance with dots in it and lines that divide the substance into rectangular shapes. What are:
a) the dots
b) the lines
c) the coloured substance?

11 How does the cell membrane protect the cell?

Cytoplasm

Cytoplasm is a watery jelly that fills most of each animal cell. It can move around inside the cell. The cytoplasm may contain stored food in the form of grains. Most of the chemical reactions that keep the cell alive take place in the cytoplasm.

Cell membrane

The **cell membrane** covers the outside of the cell. It has tiny holes in it called pores that control the movement of chemicals in or out of the cell. Dissolved substances such as food, oxygen and carbon dioxide can pass through the cell membrane. Some harmful chemicals are stopped from entering the cell by the membrane.

Parts found only in plant cells

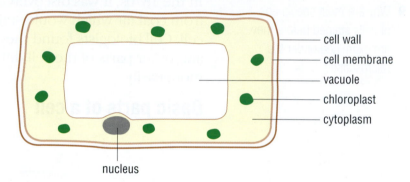

Figure 3.4 A typical plant cell

Cell wall

The **cell wall** is found outside the membrane of a plant cell. It is made of **cellulose**, which is a tough material that gives support to the cell.

Chloroplasts

Chloroplasts are found in the cytoplasm of many plant cells. They contain a green pigment, called **chlorophyll**, which traps a small amount of the energy in sunlight. This energy is used by the plant to make food in a process called photosynthesis. (You will learn more about photosynthesis later in the course.) Chloroplasts are found in many leaf cells and in the stem cells of some plants.

Large vacuole

The **vacuole** is a large space in the cytoplasm of a plant
cell that is filled with a liquid called cell sap containing
dissolved sugars and salts. When the vacuole is full of cell
sap the liquid pushes outwards on the cell wall and gives
it support. If the plant is short of water, the support is lost
and the plant wilts.

Some animal cells and **Protoctista** (page 60) have
vacuoles but they are much smaller than in plant cells.

Adaptation in cells

The word **adaptation** means the change of an existing
design for a particular task (see also page 75). You learnt
about the basic structures of animal and plant cells in the
last section, but many cells are adapted, which allows them
to perform a more specific task. Here are some common
examples of the different types of animal and plant cells.

Red blood cells

Red blood cells are disc-shaped but their centres dip
inwards. The structure is called a biconcave disc.
Red cells only have a nucleus when they are growing.
They lose it so that they can become packed with
haemoglobin when they are fully grown. **Haemoglobin**
combines with oxygen in the lungs to form oxyhaemoglobin.
The red blood cells carry this substance to parts of the
body where oxygen is needed. When the cells reach their
destination, they release the oxygen. Haemoglobin forms
again, ready for another trip to the lungs.

Figure 3.5 Red blood cell

White blood cells

White blood cells have an irregular shape. In fact, they
keep changing shape as the cytoplasm flows about inside
them. One kind of white cell, a lymphocyte, produces
antibodies, which attack harmful microorganisms in the
blood, such as bacteria. A second kind of white cell, a
phagocyte, eats the harmful microorganisms.

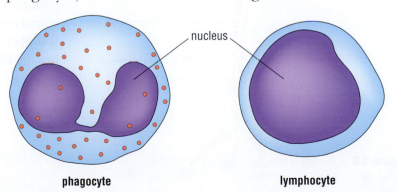

Figure 3.6 White blood cells

phagocyte   lymphocyte

Smooth muscle cells

Once you swallow food, it is moved through your alimentary canal by smooth muscle cells. These cells are spindle-shaped and lie together forming muscular tissue around the wall of the oesophagus, stomach and intestines. Muscle cells can only use their energy to contract or get shorter; they need other muscles to stretch them back to their original length. For this reason, smooth muscle cells are arranged in layers at right angles to each other. When the cells in one layer contract, they squeeze food through your body. When the cells in the next layer contract, they stretch the muscles in the first layer so they can contract again and move more food.

16 What happens to the smooth muscle cells in the outer layer in Figure 3.7 when the muscle cells in the inner layer are contracting just behind the ball of food?

Figure 3.7 The arrangement of smooth muscles in the oesophagus

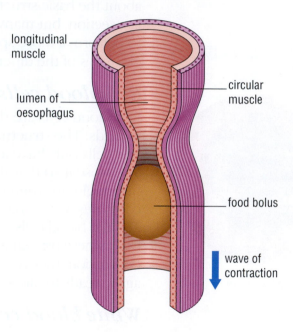

longitudinal muscle

lumen of oesophagus

circular muscle

food bolus

wave of contraction

Nerve cells

Nerves are made from nerve cells or **neurones**, which have long, thread-like extensions. These nerve cells are connected to other nerve cells in the spinal cord. The nerve cells in the spinal cord are then connected to nerve cells in the brain.

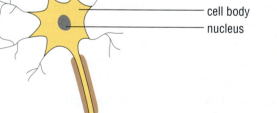

cell body

nucleus

nerve fibre

Figure 3.8
Nerve cell

Ciliated epithelial cells

Cells that line the surface of structures are called **epithelial cells**. **Cilia** are microscopic hair-like extensions of the cytoplasm. If cells have one surface covered in cilia, they are described as ciliated. Ciliated epithelial cells line the throat. Air entering the throat contains dust that becomes trapped in the mucus of the throat lining. The cilia wave to and fro and carry the dust trapped in the mucus away from the lungs.

17 Smoking damages the cilia lining the breathing tubes. What effect might this have on breathing?

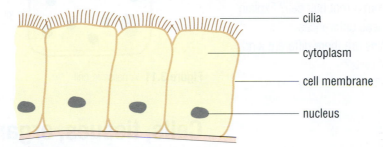

Figure 3.9 Ciliated epithelial cells

Root hair cells

Root hair cells are plant cells that grow a short distance behind the root tip. The cells have long, thin extensions that allow them to grow easily between the soil particles. The shape of these extensions gives the root hair cells a large surface area through which water can be taken up from the soil.

18 What changes have taken place in the basic plant cell to produce a root hair cell?
19 Why would it be a problem if root hair cell extensions were short and stubby?

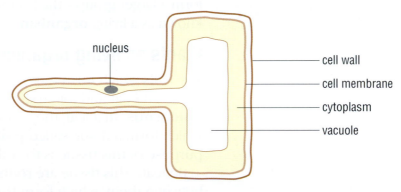

Figure 3.10 A root hair cell

Palisade cells

Palisade cells (Figure 3.11) have a shape that allows them to pack closely together in the upper part of a leaf, near the light. They have large numbers of chloroplasts in them to trap as much light energy as possible.

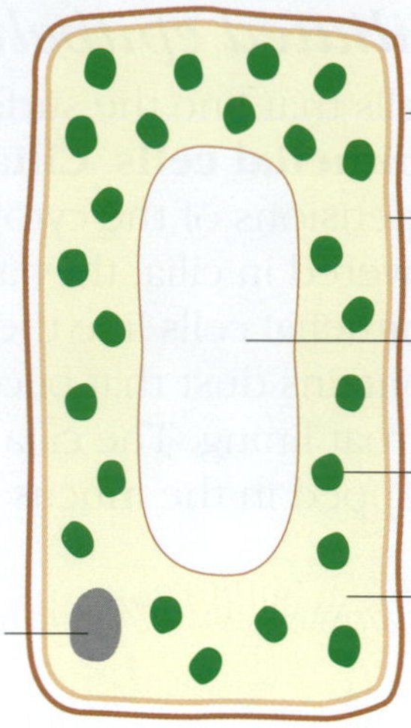

Figure 3.11 A palisade cell

Labels on figure: cell wall, cell membrane, vacuole, chloroplast, cytoplasm, nucleus

20 How is a palisade cell different from a root hair cell? Explain these differences.

21 Why are there different kinds of cells?

Cells, tissues, organs and organisms

Cells in a living thing are arranged into groups. The cells in a group are all the same kind and perform a special task in the life of the organism. This group of cells is called a **tissue**. Different tissues join together to make a larger group of cells called an **organ**. All the special tasks performed by the cells in the different tissues in the organ help the organ to keep the body alive. Organs can form groups called **organ systems**. The organs in a system perform a vital task in the survival of the body – related to the seven life processes (see page 21). All the organs and organ systems in a living thing form a larger group – the body of the living thing – which is known as a living **organism**.

Plants as living organisms

The cells which form the outer covering of a plant are broad and flat and join together to form a tissue called the **epidermis**. In a leaf, they cover a layer of palisade cells, which form a tissue called **palisade mesophyll**. The purpose of this tissue is to collect light and make food.

Beneath this tissue are rounder-shaped cells with gaps between them, which form the **spongy mesophyll** tissue. Their task is to help bring water to the leaf for use in making food. Water evaporates from their surfaces and is replaced by water drawn up in tube-like **xylem tissue**, which forms much of the mid rib and veins in the leaf.

Specialised pairs of cells in the lower epidermis of a leaf can bend to make an opening through which water vapour can escape from the spongy mesophyll. These openings are called stomata (*singular*: **stoma**).

Figure 3.12 This section through a leaf shows the different tissues that work together. The cells in the epidermis form a protective surface like tiles on a roof.

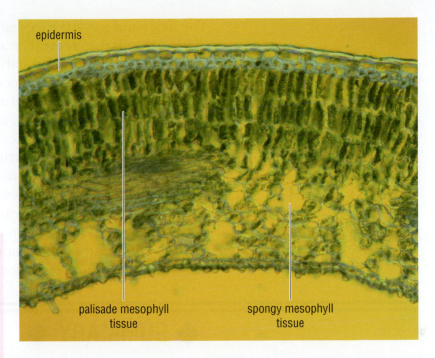

epidermis

palisade mesophyll tissue

spongy mesophyll tissue

22 Make a chart to show how different types of plant cells form tissues, which in turn form an organ. You could make a table of three columns with the headings Cells, Tissues and Organ to help you and make one or more drawings in each column. Give your chart a title, such as 'The tissues of a leaf'.

The tissues in the leaf work together and form an organ – the leaf – which makes food for the plant. Epidermis and xylem tissues join with other tissues to make other organs of the plant such as the root, stem, bud and flower. Together all the organs make up the organism – the plant.

Animals as living organisms

There are many kinds of cells in animals' bodies and they form many kinds of tissues. These in turn form groups to make organs, systems and the organism itself – the animal.

For example, the stomach is an organ that stores and digests food. It can be thought of as a muscular bag, which churns up food in the digestion process, but it is made of more than just muscle cells.

Figure 3.13 The stomach wall is made of many different tissues of cells.

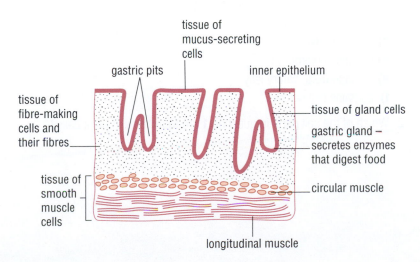

tissue of mucus-secreting cells

gastric pits

inner epithelium

tissue of fibre-making cells and their fibres

tissue of gland cells

gastric gland – secretes enzymes that digest food

tissue of smooth muscle cells

circular muscle

longitudinal muscle

The inner surface of the stomach is lined with a tissue of epithelial cells that **secrete** mucus, which helps the food slide by. There are cavities in the lining called gastric pits where tissues of gland cells secrete digestive juices to break down the food. The epithelial and gland cells are supported by a layer of fibres made from fibre-making cells. This layer connects to the layers of smooth muscle tissue.

All these tissues form the structure called the stomach, and this in turn is connected to the other digestive organs (see page 40) to form the digestive system. This system is very closely linked to other systems such as the circulatory system, where blood vessels in the intestine take away absorbed food and carry it around the body. The group of closely linked organ systems form the organism – the animal.

23 What is the connection between a cell and an organism?

◆ SUMMARY ◆

◆ The bodies of plants and animals are made of cells. The basic parts of the cell are the nucleus, cytoplasm and cell membrane (*see page 49*).

◆ In a plant cell, there is a cellulose cell wall and a vacuole (*see page 50*).

◆ Cells have different forms for different functions. They are adapted to do specific tasks in the body and life of the organism (*see page 51*).

◆ Cells are grouped together into tissues. Tissues are grouped together to make organs (*see page 54*).

◆ A number of organs that work together is called an organ system (*see page 54*).

End of chapter questions

1 Which part of a cell:
   a) has pores that control the movement of chemicals
   b) is made from cellulose
   c) contains DNA
   d) contains chlorophyll
   e) is made from watery jelly
   f) contains cell sap?

2 Which of the cell parts in question **1** are found only in plant cells?

- The fungi kingdom
- The Monera kingdom
- The Protoctista kingdom
- Viruses
- Decomposers
- Louis Pasteur

Some organisms are not made from tissues, organs and organ systems as described in Chapter 3. These organisms have a body made from only one cell and are called **microorganisms**. Most can only be seen by using a microscope but some grow together in such large numbers that their colonies can be seen by the naked eye. They are divided into three large groups or **kingdoms**. They are the fungi, Monera and Protoctista.

The fungi kingdom

It may seem surprising to find the fungi described as microorganisms, as a large number, such as the mushroom, can be easily seen without a microscope.

Figure 4.1 This is a 'fairy ring' of mushrooms. The fruiting bodies have been made at the edges of a disc-shaped mycelium in the soil.

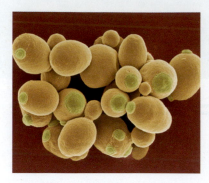

Figure 4.2 Budding yeast as seen under the electron microscope

However, at one stage in their life cycle they cannot be seen and behave like other microorganisms. This is the spore stage, which occurs when fungi reproduce. Fungal **spores** are carried by air and water currents and can travel long distances. Like other microorganisms, the spores have a protective coat that helps them survive in heat and in cold and dry conditions.

Only when the spore lands in conditions that are favourable to growth of the fungus does the spore case break open. When this happens, a thin thread called a **hypha** grows out and begins to digest any food around it. As it feeds, it grows and many hyphae are produced. As they grow, they form a larger structure called a **mycelium**, which in time produces fruiting bodies, of which the mushroom is an example. Moulds produce a mycelium that can be seen growing on food like bread, and their fruiting bodies are tiny black globes called **sporangia**.

Yeasts are microorganisms in all stages of their life cycle. They do not produce hyphae, and when fully grown they produce copies of themselves that stick to their bodies and are described as buds.

A few fungi are **parasites**. Athlete's foot is caused by a fungus that feeds on damp skin between the toes.

Useful yeast

Bread is made by mixing flour, water, yeast and sugar into a grey-white lump called dough. Inside the dough the yeast respires and produces bubbles of carbon dioxide that make the dough rise. In a bakery, the dough is cut

Figure 4.3 Dough is rising to produce the spongy texture of bread.

1 How does yeast make the spongy texture of the bread?

2 Why does the dough rise even more in a warm oven?

into pieces to make loaves. Each piece of dough is put into a baking tin. When bubbles of gas are heated they expand, so the tins are kept in a warm cupboard for about half an hour to allow the dough to rise even more. They are then placed in an oven for baking.

The Monera kingdom

The living things in the Monera kingdom have microscopic bodies made from just one cell, but the cell does not have a nucleus. There are two groups in this kingdom. They are the bacteria and the blue-green algae.

Bacteria

Bacteria are found almost everywhere – in air and water, on the surfaces of plants, animals and rocks, and inside living things too. They have spherical, spiral or rod-like shapes.

Bacteria usually reproduce by a process called **fission**, where each bacterium divides in two. If they have enough warmth, moisture and food, some bacteria can reproduce by fission once every 20 minutes. When conditions become dry and hot and unsuitable for feeding and breeding, some bacteria form spores. They can survive inside spores for a long time. They break out of the spores when favourable conditions return.

Some bacteria feed on the insides of living bodies, where they cause disease. Diphtheria, whooping cough, cholera, typhoid, tuberculosis and food poisoning are all diseases caused by different kinds of bacteria.

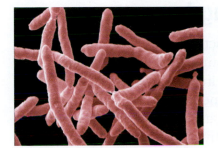

Figure 4.4 Rod-shaped bacteria as seen with an electron microscope.

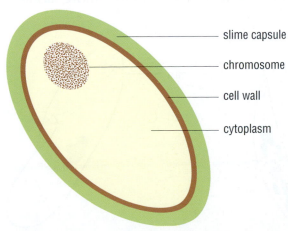
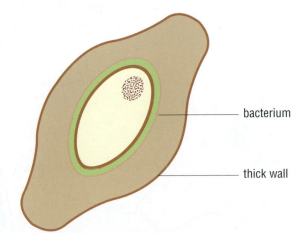

slime capsule
chromosome
cell wall
cytoplasm
bacterium
thick wall

Figure 4.5 A bacterium (left) and a bacterial spore (right)

3 If a bacterium could divide into two every 20 minutes, how many bacteria would be produced after eight hours?

Useful bacteria

Some kinds of bacteria are useful. Yoghurt is made by introducing certain bacteria to milk and making it turn sour. Vinegar is made by allowing other bacteria to feed on ethanol and change it into acetic acid.

Blue-green algae

The blue-green algae live in seas, oceans and lakes. They also grow on wet rocks at the sides of stream and rivers, at the tops of rocky seashores, and may occur widely in the soil. When they occur in huge numbers you can see them.

The Protoctista kingdom

The living things in the Protoctista kingdom have microscopic bodies made from just one cell, which contains a nucleus. They live in aquatic habitats and damp places such as soil. Some make their own food by **photosynthesis**, like plants, and are sometimes called unicellular 'plants' or protophyta. Other members of this group feed as animals do and are sometimes called unicellular 'animals' or protozoa.

The outside of the body may have a single long hair or **flagellum** (in certain kinds of Protoctista called flagellates) or a coating of tiny hairs called **cilia** (in kinds called ciliates) to help it move through the water around it. Flagellates lash the long hair like a whip to move, and ciliates wave their tiny hairs to and fro. One kind of protozoan known as *Amoeba* has neither a flagellum nor cilia but makes projections from its body wall called **pseudopodia**, or 'false feet', which help it move along and find food.

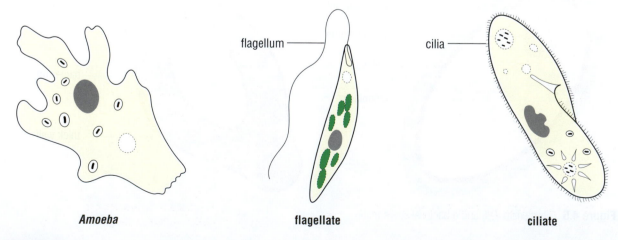

Amoeba   flagellum   cilia   **flagellate**   **ciliate**

Figure 4.6 Three members of the Protoctista kingdom

of the Monera similar to
members of the Protoctista?
b) In what ways are members
of the Monera different from
members of the Protoctista?
5 Describe the ways in which
Protoctista move.

6 Are viruses living things?
Explain your answer.
7 Produce a table of diseases
caused by viruses and bacteria.

Some members of the Protoctista live in the bodies
of other organisms, and some that live in humans
cause diseases such as malaria, sleeping sickness and
amoebic dysentery.

Viruses

Viruses do not have a cell structure. They can be stored
like mineral specimens for many years without changing.
During this time they do not feed, respire or excrete.

When they are placed on living tissues, they enter the
living cells and reproduce. They destroy the cells in the
process and may cause disease. Each kind of virus attacks
certain cells in the body. For example, the cold virus attacks
the cells that line the inside of the nose. The destructive
action of the cold virus on the cells in the nose makes the
nose run. In addition to the common cold, different viruses
also cause influenza, chicken pox, measles and rabies, and
can lead to the development of AIDS.

The virus sticks to the
membrane of a cell

The virus enters the cell, the protein coat breaks
down and releases the DNA

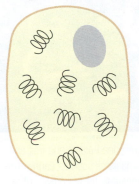

The viral DNA divides and directs the cell to make new
protein coats

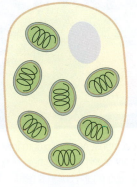

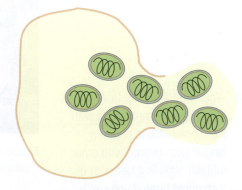

The cell wall breaks down to
release the new viruses

Figure 4.7 How a virus reproduces

Decomposers

Although people tend to think of microorganisms as harmful, the majority of them are not, and indeed they play a very important part in ensuring that life on Earth continues.

When plants and animals die, their bodies would remain unchanged if there were no bacteria and fungi. These microorganisms feed on the dead bodies and rot them down. **Minerals** from the body tissues are released into the soil in this process and become available for plants to take them up in their roots. The plants build the minerals into their own bodies as they grow and perhaps in time they pass into animals' bodies when the plants are eaten. Here the minerals are used to build up the animal's body as it grows, or take part in processes that keep the animal alive. Without microorganisms the minerals in plants and animals would be locked up in their bodies when they died and would not become available for new living things. Microorganisms ensure that the minerals that living bodies need are recycled.

8 When fruit and vegetable waste from meals is thrown out with the rubbish it ends up in a landfill site covered with other rubbish. Why is it better to use a compost heap if you can?

Figure 4.8 In a compost heap, microorganisms break down dead plants and fruit and vegetable waste from meals, with the help of some other organisms such as earthworms. The compost is spread on the soil in a garden to provide materials for the growth of more plants.

Louis Pasteur

Louis Pasteur (1822–1895) was a French scientist who made many discoveries about microorganisms. When he began his work, many people believed that living things could spring up from non-living materials. Their belief was based on observations they had made, such as that maggots appeared in rotten meat. This idea was called spontaneous generation. It was also used to explain that some forms of this emerging new life made liquids like wine and broth turn bad.

Pasteur investigated how broths turned bad in the following way. As he knew that excess heat could kill living things, he boiled some broth in flasks to kill anything that might be living in it at the start. He then heated the necks of the glass flasks until they were soft, and pulled them out into a long, thin, curving tube called a swan-neck. The broths in the flasks did not go bad.

Figure A Louis Pasteur at work in his laboratory

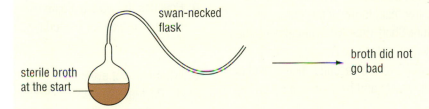

Figure B The first stage of Pasteur's swan-necked flask experiment

Then Pasteur broke open one of the flasks and exposed the broth to the open air. The broth went bad. This made Pasteur conclude that there was something in the air that was making the broth go bad. In the swan-necked flasks where the broth did not go bad, this 'something' might have settled in the bend of the neck and therefore not reached the broth. To test his idea he tipped a swan-necked flask so that some of the broth went into the bend where dust and other particles may have collected, and then he tipped it back again. The broth in this second flask then went bad.

1 In Pasteur's first investigation, how did he make his test fair?
2 Gases mix freely in the air. How did Pasteur's first experiment show that it was not a gas that caused the broth to go bad?

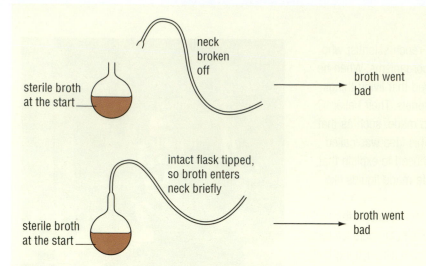

sterile broth at the start

neck broken off

broth went bad

intact flask tipped, so broth enters neck briefly

sterile broth at the start

broth went bad

Figure C Further stages of Pasteur's swan-necked flask experiment.

Pasteur concluded that whatever was causing the change could be carried by air currents, but it must be heavier than air as it settled in the bend in the swan neck. He described these tiny objects as spores.

Pasteur went on to show that the souring of wine and milk was due to microorganisms and not spontaneous generation. He developed a process to help stop these liquids going bad quickly, which was named after him – pasteurisation. In one of the main forms of pasteurisation used today on milk, called High Temperature/Short Time (HTST) treatment, the milk passes through pipes surrounded by hot water. During this time, the temperature of the milk is raised to 72°C for 16 seconds, then cooled to 4°C. This process kills mould spores, yeasts and bacteria that cause the milk to turn sour.

After realising that broths were made bad by microorganisms, Pasteur reasoned that microorganisms might cause disease in animals and in humans. When silk farmers asked him to explain why their insects were dying, Pasteur performed investigations which showed that the disease was due to microorganisms. He later went on to discover that chicken cholera was due to a microorganism attack and found that anthrax, a disease of cattle and humans, was also due to microorganisms.

This last discovery showed that there was a link between microorganisms and human diseases. Pasteur believed in and gave great support to a developing **theory** called the Germ Theory. This says that many diseases are caused by microorganisms invading the body. Pasteur's work led many scientists from then on to work to discover the causes of diseases and how to prevent them.

3 How do you think Pasteur's experiments with broths affected the way people thought about spontaneous generation?

4 How has the work of Pasteur helped the way we live today?

5 What piece of evidence did Pasteur use when he decided to boil the broths?

6 What prediction do you think Pasteur might have made before breaking open one of the flasks?

7 What did Pasteur use to think of the idea that microorganisms might cause disease in animals and humans?

8 What evidence produced by Pasteur supports the Germ Theory?

◆ SUMMARY ◆

◆ Mushrooms, moulds and yeast are members of the fungi kingdom (*see page 57*).

◆ Athlete's foot is a disease caused by a fungus (*see page 58*).

◆ Bacteria and blue-green algae are members of the Monera kingdom (*see page 59*).

◆ Some bacteria cause diseases (*see page 59*).

◆ The protozoa and protophyta are members of the Protoctista kingdom (*see page 60*).

◆ Some protozoa cause diseases (*see page 61*).

◆ Viruses cause diseases (*see page 61*).

◆ Microorganisms decompose dead plants and animals (*see page 62*).

◆ Louis Pasteur showed that microorganisms cause disease (*see page 63*).

End of chapter question

Yeast feeds on sugar when the sugar is dissolved in water. As it feeds, it respires and produces carbon dioxide gas, which makes bubbles that form froth on the water surface. Using this information, plan an investigation to find out if temperature affects the rate at which yeast feeds on sugar.

◆ Key words in ecology
◆ Food chains
◆ Investigating a habitat
◆ Adapting to daily changes
◆ Adapting to seasonal changes
◆ Adapting to certain habitats
◆ Adaptations for feeding

Ecology

What is the countryside like where you live? Are there trees or bushes? Maybe there are both or maybe neither, with the land covered with grass and other small plants. Are there woodlands or perhaps a forest? Can you see a stream, a river, a pond or a lake? Or maybe you live near the seashore. What about animals that you can see easily? The chances are you can see a few different kinds of birds and insects and occasionally a squirrel, deer or monkey, while under stones there may be slugs, snails and spiders. The scene may look too complicated to investigate scientifically, but the study of **ecology** – the study of living things in their environment – was established at the beginning of the 20th century to do just this.

Key words in ecology

There are three words often used in the study of ecology. They are habitat, environment and ecosystem. Here are the meanings of these terms.

Habitat

The **habitat** is the home of a plant or animal – the place where it lives.

A large number of different living things can share the same habitat. For example, in Europe deer, woodpeckers, oak trees and bluebells share the same habitat – the woodland. All the plants and animals living in a habitat form a **community** of living things.

1 What is in the community of living things in a pond habitat?

Figure 5.1 Sea anemones (red), whelks (white) and cone-shaped limpets live in rock pools in a seashore habitat.

Environment

The **environment** of a living thing is everything around it that could affect its survival. The things that affect plants and animals are called environmental factors. There are factors due to the non-living part of the environment, which include temperature, wind speed, the amount of light and the amount of water available. These are called **abiotic factors**. There are factors due to the living things in the environment, such as the amount of food or the number of predators. These are called **biotic factors**.

Ecosystem

The **ecosystem** is made up from the community of living things and the abiotic factors. The way these two parts of the ecosystem react together allows the community to survive for a very long time – until something upsets the balance between the two parts. The ecosystem only needs energy, almost always from sunlight, to keep it self-sustaining. Some ways in which the parts of a very simple ecosystem react together are shown in Figure 5.2.

Food chains

In the diagram of the ecosystem in Figure 5.2, you can see how the plant and the animal are linked together by the passage of food between them. This is shown by the arrow going from the food (the plant) to the feeder (the animal). In a **food chain**, the organism that makes the food

2 a) What environmental factors might affect an earthworm living in a lawn?
   b) Which ones are abiotic factors and which are biotic factors?
3 What environmental factors might affect a plant growing on a mountain top?

4 The Sun provides another abiotic factor in the ecosystem. What is it and how is it measured?
5 Name another abiotic factor, besides those provided by the Sun, which can affect the growth of the plants.
6 Name a biotic factor that could affect the number of rabbits in the ecosystem.

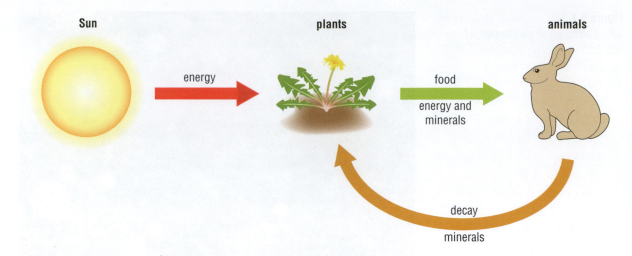

Figure 5.2 The arrows show the passage of energy and minerals through an ecosystem.

(the plant), is called the **producer** and the organism that eats the food is called the **consumer**. Most food chains are much longer than this. For example, in a European woodland habitat woodmice eat plants and tawny owls eat mice. This can be represented by the food chain:

plant → wood mouse → tawny owl

The plant is the producer. The wood mouse is the **primary consumer** (the first consumer of the food made by the plant) and is also called a **herbivore**. The tawny owl is the **secondary consumer** (the second consumer of the food originally made by the plant) and is called a **carnivore**. You can see a tawny owl attacking a wood mouse in Figure 5.12 (page 77).

Food chains may be even longer. For example, in many warmer parts of the world grasshoppers eat plants, frogs eat grasshoppers, snakes eat frogs and hawks eat snakes. This can be represented by the food chain:

plant → grasshopper → frog → snake → hawk

Again, the plant is the producer but this time the grasshopper is the primary consumer and a herbivore, the frog is the secondary consumer and a carnivore, the snake is the third consumer, called the **tertiary consumer** (and also a carnivore) and the hawk is the fourth consumer, called the **quaternary consumer** (and also another carnivore). The highest-level consumer in a food chain is called the top carnivore.

7 In this chapter, you will read about the following animals: roe deer, stoat, ptarmigan, lung fish, swallow, mountain goat, fur seal, cuttlefish. For each one, construct a food chain from the information given. Say what kind of consumer the animal is, whether it is a herbivore or a carnivore, and whether it is prey and/or predator or neither. You may like to answer this question now or as you reach each animal in the chapter.

a) roe deer – fox feeds on roe deer, roe deer feeds on grass

b) stoat – rabbit feeds on grass, stoat feeds on rabbits

c) ptarmigan – eagle feeds on ptarmigan, ptarmigan feeds on plants

d) lung fish – lung fish feeds on snails, snail feeds on plants, crocodile feeds on lung fish

e) swallow – insect feeds on plants, sparrowhawks feed on swallow, swallow feeds on insects

f) mountain goat – wolf feeds on mountain goat, mountain goat feeds on plants

g) fur seal – shrimps feed on algae in plankton, killer whale feeds on fur seal, fur seal feeds on shrimps

h) cuttlefish – crab feeds on sea snail, dolphin feeds on cuttlefish, sea snail feeds on algae, cuttlefish feeds on crab

Some animals, such as bears, feed on both plants and animals. They are called **omnivores**. An omnivore feeds as a primary consumer when it feeds on plants, and as a secondary or higher-level consumer when it feeds on animals.

Figure 5.3 This snake is a predator, about to attack its prey.

Biodiversity

Biodiversity is the term used to describe the number and variety of species in an ecosystem. Most ecosystems have a high biodiversity, which means they have a large number of different species showing a great variety of features. If the balance between some species in the ecosystem or between the abiotic factors and the community is destroyed, the biodiversity may fall as some plants and animals die out.

Investigating a habitat

A habitat, such as a wood or a pond, is a home to a wide range of living things. It has a high biodiversity. When ecologists study a habitat they need to collect data about it. This not only provides information about the organisms that live there, but the data can also be stored and used to **monitor** the habitat – at a later date, another survey can be made and the data obtained can be compared with the previous data. This shows how the **populations** of species have fared in the time between the surveys. Some may have increased, others decreased and some stayed the same. By comparing data in this way, the biodiversity of the habitat can be monitored. Then, if it falls, events close by, such as a change in land use or the release of pollutants, can be studied carefully for signs of environmental damage.

Recording the plant life in a habitat

When a habitat is chosen for study, it must be examined closely to collect data. Sometimes cameras can be used to gather photographic evidence for later study. Often a map is made in which the habitat boundaries and major features, such as roads or cliffs, are recorded. The main kinds of plants growing in the habitat are identified and the way they are distributed in the habitat is recorded on the map.

A more detailed study of the way the plants are distributed can be made by using a quadrat and by making a transect.

Using a quadrat

A **quadrat** is a square frame. It is placed over an area of ground and the plants inside the frame are recorded. When using a quadrat, the area of ground should *not* be chosen carefully. A carefully selected area might not give a fair record of the plant life in the habitat, but may support an idea that the ecologists have worked out beforehand. To make the test fair, the quadrat is thrown over the shoulder so that it will land at random. The plants inside it are then recorded. This method is repeated a number of

Figure 5.4 Using a quadrat to map the daisies in a lawn

8 How could you use a quadrat to see how the plants in a particular area change over a year?

times and the results of the random samples are used to build up a record of how the plants are distributed. An estimate of how many of each kind of plant there are in the area can then be made.

Making a transect

If there is a feature such as a bank, a footpath or a hedge in the habitat, the way it affects plant life is investigated by using a **line transect**. The position of the transect is chosen carefully so that it cuts across the feature being examined. The transect is made by stretching a length of rope along the line to be examined and recording the plants growing at certain intervals (called stations) along the rope. When plants are being recorded along a transect, abiotic factors such as temperature or dampness of the soil may also be recorded to see if there is a pattern between the way the plants are distributed and the varying **abiotic** factors.

Figure 5.5 Making a transect down a hillside

Collecting small animals

Different species of small animals live in different parts of a habitat. In a land habitat they can be found in the soil, on the soil surface and leaf litter, among the leaf blades and flower stalks of herbaceous plants, and on the branches, twigs and leaves of woody plants. They can be collected from each of these regions using special techniques.

Collecting from soil and leaf litter

A Tullgren funnel is used to collect small animals from a sample of soil or leaf litter (Figure 5.6). The sample is placed on a gauze above the funnel and a beaker of water is placed below the funnel. The lamp is lowered over the sample and switched on. The heat from the lamp dries the soil or leaf litter and the animals move downwards to the more moist regions below. Finally, the animals move out of the sample and into the funnel. The sides of the funnel are smooth so the animals cannot grip onto them and they fall into the water.

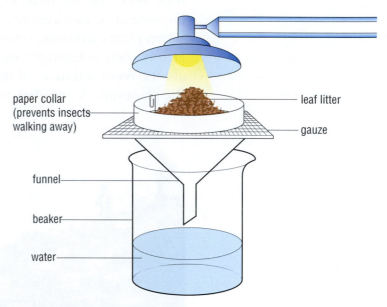

paper collar (prevents insects walking away)

leaf litter

gauze

funnel

beaker

water

Figure 5.6 A simple Tullgren funnel

Pitfall trap

The pitfall trap is used to collect small animals that move over the surface of the ground. A hole is dug in the soil to hold two containers, such as yoghurt pots, arranged one

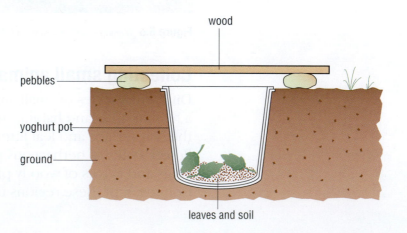

wood

pebbles

yoghurt pot

ground

leaves and soil

Figure 5.7 A pitfall trap

9 What is the advantage of using an outer and an inner container instead of just one container for the pitfall trap?

10 Why are large leaves not used inside the trap?

inside the other. The containers are placed in the hole, and the gap around them up to the rim of the outer container is filled in with soil. A few small leaves are placed in the bottom of the container and four pebbles are placed in a square around the top of the trap. A piece of wood is put over the trap, resting on the pebbles. The wood makes a roof to keep the rain out and hides the container from predators. When a small animal falls in, it cannot climb the smooth walls of the inner container and remains in the trap, hiding under the leaves until the trap is emptied.

Sweep net

The sweep net is used to collect small animals from the leaves and flower stems of herbaceous plants, especially grasses. The lower edge of the net should be held slightly forward of the upper edge to scoop up the animals as the net is swept through the plants. After one or two sweeps the mouth of the net should be closed by hand and the contents emptied into a large plastic jar where the animals can be identified.

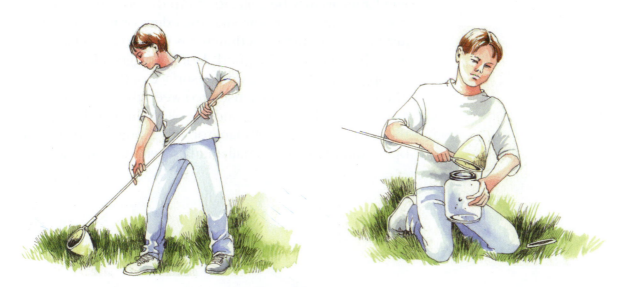

Figure 5.8 Using a sweep net

Sheet and beater

Small animals in a bush or tree can be collected by setting a sheet below the branches and then shaking or beating the branches with a stick. The vibrations dislodge the animals, which then fall onto the sheet. The smallest animals can be collected in a pooter (Figure 5.9).

Pooter

11	What is the purpose of the cloth cover on the end of tube B inside the pooter?

Tube A of the pooter is placed close to the animal and air is sucked out of tube B. This creates low air pressure in the pooter so that air rushes in through tube A, carrying the small animal with it.

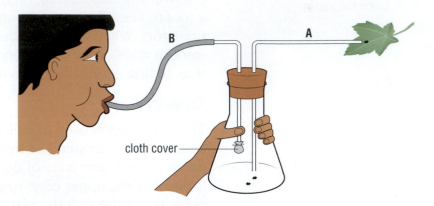

Figure 5.9 Using a pooter

Collecting pond animals

Pond animals may be collected from the bottom of the pond, the water plants around the edges or the open water just below the surface. A drag net is used to collect animals from the bottom of the pond. As it is dragged along, it scoops up animals living on the surface of the mud.
The pond-dipping net is used to sweep through vegetation around the edge of the pond to collect animals living on the leaves and stems. A plankton net is pulled through the open water to collect small animals swimming there.

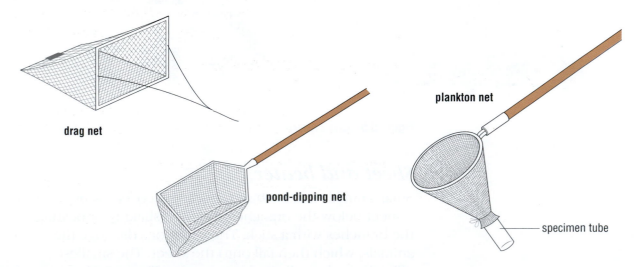

Figure 5.10 Three types of pond net

Studying the seashore

There are two kinds of seashore – sandy shores and rocky shores. A transect can be made from the top of each type of shore down to the water's edge at low tide. On the sandy shore, a quadrat can be placed on the sand at 10-metre intervals down to the sea and any living things on the surface recorded. The quadrat can then be removed and the sand dug out to a depth of 50 cm and sifted through a sieve to collect any small animals. During the digging, larger animals such as worms and molluscs may be encountered as the animals burrow away from the disturbance. A note of their presence could be made. After the specimens have been extracted and examined, they must be returned to their habitat.

On a rocky shore, a transect can be made with a rope, and the presence and identity of seaweeds can be noted at 10-metre intervals along it. Small rock pools may have a quadrat placed over them and their outline marked on squared paper. The positions of seaweeds and animals such as limpets, mussels and sea anemones could be plotted inside the outline. The temperature of rock pools could be taken at 10-metre intervals along the transect.

Adaptations

When we adapt to something we adjust to it. For example, if you change schools you may find that the new school runs in a slightly different way to the old one and in time you adjust or adapt to it. In a similar way, plant and animal species are adapted to their habitats. The features they have that help them survive there are called **adaptations**.

Adaptations to changing conditions

As the Earth spins on its axis, it brings each place on its surface towards the Sun and then away from it, in a daily cycle. The part of the Earth that is facing the Sun at any time is in daylight. Then as the Earth turns, this part faces away from the Sun, and it becomes night. The Earth continues to turn, and this part comes back into daylight once more. Some plants and animals have special adaptations to these daily changes. The examples given below are mainly found in northern Europe.

Plant adaptations to daily changes

Figure 5.11 Crocus flowers opening in the day in a woodland habitat.

Some plants, such as the crocus and the tulip, open their flowers during the day and close them at night. The flowers are open in the day so that insects may visit them for nectar, and in return transport pollen from one flower to another, bringing about pollination. The flowers close at night to protect the delicate structures inside the petals from low temperatures and from dew. The dew could wash the pollen off the stamens (male parts of the flower) so that it cannot be picked up and transported by the insects.

The night-scented stock is an unusual plant in that its flowers open in the evening and close during the day. This adaptation allows moths to visit the flowers and pollinate them.

A few plants make movements of their leaves over a 24-hour period. Leaves not only make food but they also provide a large surface area for the evaporation of water, which in turn helps to draw water through the plant from the roots. In wood sorrel, for example, each leaf is divided into three leaflets. During the day the leaflets spread out and become horizontal. In this position, they are best placed for receiving sunlight to make food, and for losing water to the air, which causes the plant to draw up water from the roots. The water is needed to make food, and it also helps to keep the leaves cool in the sunlight. In the evening, the leaflets fold close together. This helps them to lose less water when the plant is not making food.

Plants can make these movements by moving water about inside their bodies. When there is a lot of water in one place it makes the **cells** swell up. When some of it is removed from a place the cells sag. This swelling and sagging of the cells allows parts of the plant to move slowly.

12 Do the petals of a crocus and the leaves of a wood sorrel plant move for the same reason? Explain your answer.

For discussion

Consider plants that grow in your local area which show these adaptations.

Animal adaptations to daily changes

Animals are adapted to being active at certain times of the day and resting at other times. At night, most birds roost (sleep) but as soon as it is light they may start flying about in search of food. They have large eyes and consequently have excellent vision, which is essential for them to fly, land and search for food.

Figure 5.12 A tawny owl swoops down on a wood mouse in a European forest.

The tawny owl has several adaptations that allow it to catch mice at night. It has large eyes that are sensitive to the low intensity of light in the countryside at night. These allow it to see well enough to fly safely. The edges of some of the owl's wing feathers are shaped to move noiselessly through the air when the bird beats its wings. This prevents the mouse's keen sense of hearing from detecting the owl approaching in flight.

The owl has sharp talons on its toes that act as daggers, to kill its prey quickly and to help carry the prey away to be eaten at a safe perch.

At night, most birds are replaced in the air by bats. These animals roost in the day and come out at dusk to feed. Many species hunt for flying insects. Bats do not use their eyes but have developed an echo-location system. They send out very high-pitched squeaks that we cannot hear. These sounds reflect off all the surfaces around the bat and travel back to the bat's ears. The bat uses the information from these sounds to work out the distance, size and shape of objects around it. This allows the bat to fly safely and detect insects in the air, which it can swoop towards and eat.

Many insects, such as butterflies, bees and wasps, are active and fly during the day. At night, moths take to the air to search for food.

The squirrel is a mammal that is active during the day. Deer may also be active but they hide away in vegetation. Field mice and voles may be active during the day but hide in the grass and other low vegetation. These animals are also active for periods at night, when they are at risk of falling prey to night predators such as the fox and the owl.

13 What adaptations does the tawny owl have that allow it to detect, approach and attack its prey?

14 Why should the owl kill its prey quickly?

15 What adaptations do you think a mouse may have to help it survive a predator's attack?

16 Imagine that you are camping in a wood.
  a) What animals would you expect to see during the day?
  b) What animals are active in the evening?

While darkness is the major feature of a habitat at night, there is also a second feature – humidity. At night all surfaces in the habitat cool down and the air cools down too. This causes water vapour in the air to condense and form dew. The increase in humidity is ideal for animals such as slugs and woodlice, which have difficulty retaining water in their bodies in dry conditions. They hide away in damp places during the day but as more places in the habitat become damp at night they become more active and roam freely searching for food. In the morning as the humidity decreases they hide away again somewhere damp.

Plant adaptations to seasonal changes

The abiotic (non-living) factors in a habitat change with the seasons. The grass plant is adapted to survive winter conditions but its short roots make it dependent upon the upper regions of the soil staying damp. In drought conditions, the soil dries out and the grass dies. Daffodils are adapted to winter conditions as the leaves above the ground die and the plant forms a bulb in the soil. Bark is an adaptation of trees that provides a protective insulating layer around the woody shoot in winter.

Many trees have broad, flat leaves. They lose a great deal of water through them. In dry seasons, or in regions where the ground freezes in winter and prevents water being taken up by the roots, the trees lose their leaves and grow new ones when conditions improve. These trees are called **deciduous** trees.

Figure 5.13 Deciduous trees in a European woodland

17 In what ways are plants adapted to survive winter conditions?

Some trees, such as the holly and most conifers, have leaves that lose little water in winter. This allows the trees to stay in leaf all through the year. Trees that behave in this way are called **evergreen** trees.

Plants that float on the open water of a pond in spring and summer do not remain there in the winter. Duckweed produces individuals that sink to the pond floor, while the water plant called frogbit produces heavy seeds.
The plants around the water's edge die back and survive in the mud as thick stems called **rhizomes**.

Figure 5.14 Water lilies become dormant in winter. The foliage (leaves) dies back, and the plant survives under the surface, where the water does not freeze, until spring returns.

Animal adaptations to seasonal changes

The roe deer (Figure 5.15) lives in woodlands in Europe and Asia. In the spring and summer, when the weather is warm, it has a coat of short hair to keep it cool. In the autumn and winter, it grows longer hair that traps an insulating layer of air next to its skin. This reduces the loss of heat from its body.

The stoat, which lives in northern Europe, northern Asia and Canada, grows a white coat in the winter, which loses less heat than its darker summer coat. The stoat preys on rabbits and its white coat may also give it some camouflage when the countryside is covered in snow.

The ptarmigan is about the size of a hen. It lives in the north of Scotland, northern Europe and Canada.
In summer, it has a brown plumage that helps it hide away from predators while it nests and rears its young.
In winter, it has a white plumage that reduces the heat

18 How do the adaptations of the following animals help them to survive in the winter?
a) roe deer
b) stoat
c) ptarmigan
19 How might their winter adaptations affect their lives if they kept them through the spring and summer?

Figure 5.15 A roe deer in European grassland in summer (left) and a ptarmigan in the tundra in winter (right)

lost from its body and gives it camouflage. Feathers grow over its toes and make its feet into 'snowshoes', which allow it to walk across the snow without sinking.

The lung fish of Africa and South America live in rivers, but when the rivers dry up in the dry season they can still survive. They make a burrow in the riverbed and rest there, breathing air until the rainy season returns. This kind of rest through a hot dry season is called **aestivation**.

Figure 5.16 An African lung fish out of water. It has a pair of lungs, which it uses to extract oxygen from the air.

In woodlands in Europe, insects avoid harsh winter conditions by spending their lives in the inactive stages of their life cycles – the egg and the pupa. This means that insect-eating animals such as bats have nothing to eat in winter, so they store up fat in the autumn to give them energy to sleep though the winter in a state of **hibernation**.

eggs

larva (caterpillar)

pupa (chrysalis)

adult (butterfly)

Figure 5.17 The life cycle of a butterfly

Insect-eating birds such as the swallow leave the countries of northern Europe in the autumn and fly to Africa. They spend the winter feeding on insects in Africa, and return to Europe in the spring. When an animal moves its location as the seasons change it is said to **migrate**.

Adaptations to certain habitats

The climate and the landscape create the conditions in a habitat. Two of the more extreme habitats are deserts and mountains but some living things have adapted to the conditions there and survive.

Deserts

In deserts there may be a short rainy season followed by a long dry season. Some flowering plants have very short life cycles so that their seeds can germinate as soon as it rains. They can then grow, flower and set seed before the soil loses all of its moisture. Cacti survive dry conditions by storing water inside their bodies. They have a thick waxy covering to prevent the water from escaping from their surfaces, and spikes to prevent animals from biting into them for a drink. Some cacti have roots that spread out a long way just underneath the soil surface. They collect as much water as possible from the upper part of the soil when the rains arrive. Other cacti have long roots that grow deep into the soil to collect the water that has drained down there.

The camel has many adaptations for survival in the desert. After drinking a hundred litres of water it can walk for several days without taking another drink. Its feet have thick pads, which insulate it from the hot desert sand. The feet are also webbed so that their weight is spread out over a larger area. This reduces the pressure on the sand, and stops the camel from sinking into it. The camel has long legs, which hold the body above the hot air close to the ground. There are muscles in the camel's nose that enable it to shut its nostrils. This keeps sand out of the respiratory system during a sandstorm. The eyes have long lashes, which keep flying sand from reaching the eyes in a sandstorm. If sand does get into an eye, camels have a third eyelid to get it out. It moves from side to side and wipes the sand particles away. It is so thin that a camel can see through it, and camels often keep this third eyelid closed while walking through a sandstorm. Desert plants have tough leaves but the camel has strong teeth to grind them up. If the camel cannot find food, it uses energy stored in the fat in its hump.

For discussion

What clothes do people wear to help them survive in a desert? Explain how the clothes help the people to survive.

Figure 5.18 Cacti growing in a desert in North America

Figure 5.19 Dromedaries or Arabian camels, like this one, are found in the deserts of Africa, the Arabian Peninsula and Australia.

Mountains

The conditions on mountaintops are similar to conditions in the polar regions. There are long seasons when it is too cold for plants to grow and the ground is covered with snow, which makes it difficult for animals to move around and find food. Plants on mountains may spend the cold season as seeds and sprout into life when warm conditions arrive. The plants complete their life cycles in a few weeks so that the seeds are ready for the next season of cold weather. Some plants survive the cold conditions by having hairy leaves. These hairs prevent the plants from losing water, and trap air to provide insulation. Mountain plants also grow close to the ground; if they grew tall they would be damaged by the frequent strong winds at high altitudes.

Figure 5.20 Moss campion is growing very low on the ground on this alpine surface. The flowers are produced on the side of the moss cushion that receives the most light.

20 A seed from a plant adapted for growing in lowland meadows is carried by the wind up a mountain and lands at its top. The seed then germinates and starts to grow in the warm season. What may happen to the plant when the cold season arrives? Explain your answer.

Birds such as the golden eagle and raven visit the mountaintops in summer to look for food, but avoid them in winter. Large animals such as mountain goats and red deer also visit high mountainsides to feed on the plants, but move down the mountain before winter to avoid being trapped in the snow. The ptarmigan (page 80) is a bird that remains on the mountains throughout the year. It changes its plumage to match its surroundings and avoid predators. The mountain hare also changes its colour from brown in summer to white in winter.

Figure 5.21 The mountain goat of North America can find small spaces on rock faces in which to place its feet and keep its balance as it feeds

Aquatic habitats

Aquatic habitats are watery habitats. There are freshwater habitats, such as ponds and rivers, and salt-water habitats, such as seas and oceans and their sandy or rocky shores.

- **Freshwater plants**. The roots of land plants have oxygen around them in the air spaces in the soil. In the waterlogged mud at the bottom of a pond there is very little oxygen for the root cells. The stems of water plants such as water lilies (Figure 5.14) have cavities in them through which air can pass to the roots in order to overcome this problem.

  Water plants use the gases they produce to hold their bodies up in the water and therefore do not need strong, supporting tissues, as land plants do. Minerals can be taken in from the water through the shoot surfaces of the water plant, leaving the root to act as an anchor. The leaves of submerged water plants are thin, allowing minerals in the water to pass into them easily. The leaves also have feathery structures that make a large surface area in contact with the water. This further helps the plant to take in all the essential minerals. Floating water plants like duckweed have a root that acts as a stabiliser.

- **Freshwater animals**. Although it lives underwater, the diving beetle breathes air. It comes to the surface and pushes the tip of its abdomen out of the water. The beetle raises its wing covers and takes in air through breathing holes, called spiracles, on its back. (In insects living on land the spiracles are on the side of the body.) When the beetle lowers its wing covers, more air is trapped in the hairs between them. It is able to breathe this air while it swims underwater. Diving beetles feed on a range of foods, including small fish, tadpoles and other insects.

21 In what ways are the features of a plant living in water different from a plant living on land?

Figure 5.22 A diving beetle feeding on an earthworm

● **Marine algae**. Billions of algae live in the sunlit waters of seas and oceans. They contain drops of oil to help them float, and long spines to slow down the speed at which they sink. The more slowly they sink, the greater the chance of a water current pushing them back up near to the surface. These algae make food by **photosynthesis**, and so it is vital for them to remain in water that receives sunlight.

Seaweeds are large algae that live at the edges of seas and oceans and on rocky shores. They too must stay in sunlit water, but close to the shore there are strong currents due to the tides. The seaweeds have root-like structures called holdfasts, which grip the rocks and stop the seaweeds from being swept away (page 113).

Figure 5.23 Fur seals diving around large seaweeds called kelp

● **Marine animals**. These are animals that live in salt water. The seas and oceans are home to a great variety of animal species, and new species are discovered every year. Animals that live on the ocean floor, such as the sea spider, have long legs to help them walk over the mud. In the deep ocean water, where sunlight does not reach, many animals have special organs to generate light. Light generated by living things is called **bioluminescence**. Very little heat is generated by the chemical reactions that produce light, so the cells of the organism are not damaged. The light is used by the animals to recognise each other in the dark and to find food.

Figure 5.24 A cuttlefish with light-emitting organs

22 What might happen if living organisms that emit light energy also produced a large amount of heat energy?

Animals on the seashore are in danger of being swept away by water currents. Lugworms and molluscs, such as cockles and razor shells, burrow in sandy beaches to stay on the shore. The limpet has a fleshy foot, which acts like a sucker to hold it in place on a rocky shore (Figure 5.1). Sea anemones also have a sucker-like base that helps them to grip the sides of a rock pool.

Adaptations for feeding

There are two main ways in which animals are adapted for feeding. There are animals that are adapted for feeding on plant foods (these animals are called herbivores), and those that are adapted for feeding on other animals (carnivores).

Herbivores

When people think of herbivores, they tend to think of herbivorous mammals such as cattle or deer. Herbivores exist in other animal groups too. For example, caterpillars, which are insects, and slugs and snails, which are molluscs, are all herbivores.

23 a) Design a bark-feeding mammal that burrows its way from tree to tree.

b) What adaptations would you give it to protect it from predators?

Plant material is tough, so herbivorous animals have adaptations that allow them to break it up for digestion. Herbivorous mammals such as cattle, sheep and antelope have large, strong back teeth that they use for grinding up the food. Caterpillars have strong jaws for nibbling along the edge of a leaf, while slugs and snails have a tongue covered in tiny teeth, which they use like sandpaper to rasp away at the surface of their food.

Herbivorous animals are the prey of carnivorous animals, so they have developed features that help them reduce their chances of being caught and eaten. Many herbivores, from caterpillars to giraffes, have body colours that help them blend into their surroundings – they have **camouflage**.

Figure 5.25 Giraffes are the tallest living animals. Their colouring is very similar to their surroundings on the savannah and this helps to camouflage them

Some herbivores, such as deer (Figure 5.15, page 80), may also hide away in vegetation during the day and come out into the open at night when it is difficult for carnivores to see them.

Herbivorous mammals such as the rabbit (Figure 7.17, page 115) have eyes on the sides of their head. This gives them a very wide field of vision, enabling them to see a carnivore approaching. Rabbits, like many herbivorous animals, have large ears that can be turned to face almost every direction so that the sound of an approaching predator can be detected.

Carnivores

Just as people think of a cow as a herbivore, they may think that all carnivores are like the lion or leopard. Carnivores, like herbivores, come in all shapes and sizes. Spiders, for example, are carnivores. They set web traps to catch their prey. The frog is also a carnivore, and can flick out its tongue very quickly to catch flies. Most carnivorous mammals have large, conical canine teeth for stabbing their prey, and molars that are adapted for holding bones while the jaw muscles press on them to crack them open for their marrow. The shrew belongs to a group of mammals that feed mainly on insects, called the insectivores. Its teeth are pointed, making the jaws look like those of a miniature crocodile. This arrangement of teeth allows the shrew to catch hold of the tough body of an insect and chew it up.

Animals that catch **prey** are called **predators**. Predatory birds such as eagles, hawks and owls are known as birds of prey and are adapted for catching and feeding on other animals. They have long claws on their feet called talons, which they use to grab and stab their prey. They also have hooked beaks for ripping up their prey into smaller pieces that are easy to swallow.

Carnivorous birds and mammals share an adaptation – they both have eyes that face forwards. This means that the field of vision of each eye partly overlaps the field of vision of the other eye, and this allows the animal to judge distance. Without this overlap, judging distance can be very difficult. You can test this yourself by putting a pen and its top on the table. Close one eye, look at the two objects, and then pick them up and try quickly to put the top on the pen. The chances are that the first time you try this you will miss. Carnivorous animals need to be able to judge distance extremely accurately to pounce on their prey. If they miss, they go hungry.

As herbivorous mammals are constantly looking, listening and even sniffing the air for signs of an approaching predator, predators themselves have to take care when they are hunting. Some predators such as lions even set up an ambush to catch their prey.

24 Design a bird that feeds on the mammal you invented for question **23**. Explain the reasons for the features you give it.

25 A starling pulls up a worm and eats it. Later a sparrowhawk attacks and kills the starling and carries it away for a meal. Do these observations support the idea that carnivores are always predators? Explain your answer.

26 How does the field of vision of a herbivore compare with that of a carnivore?

Figure 5.26 Lions have forward-facing eyes and a body colour that camouflages them in the dry grass of the savannah.

◆ SUMMARY ◆

◆ Ecology is the study of living things and where they live (*see page 66*).

◆ The home of a living thing is called its habitat (*see page 66*).

◆ Food passes from one species to another along a food chain (*see page 67*).

◆ Some plants are adapted to changes that occur during the day (*see page 76*).

◆ Animals are adapted to changes that occur during the day (*see page 77*).

◆ Plants are adapted to changes that occur with the seasons (*see page 78*).

◆ Animals are adapted to changes that occur with the seasons (*see page 79*).

◆ Some plants and animals are adapted for survival in a desert (*see page 81*).

◆ Some plants and animals are adapted for survival on mountains (*see page 83*).

◆ Some plants and animals are adapted for survival in fresh water (*see page 84*).

◆ Many organisms are adapted for survival in a salt-water habitat (*see page 85*).

◆ Some animals are adapted for feeding on plants (*see page 86*).

◆ Some animals are adapted for feeding on other animals (*see page 87*).

End of chapter questions

1 How would you set about investigating a habitat such as a hedgerow?

2 One end of a transect was set up in a grassy area 5 m from a tree trunk. The other end was set up 5 m from the trunk on the other side of the tree. The tree's branches stretched to 2 m on either side of the trunk. A pitfall trap was set up at one end of the transect and at 1 m stations to the other end. After one day the pitfall traps were emptied. The table shows the results.

Pitfall trap	0	1	2	3	4	5	6	7	8	9	10
Herbivorous beetles	6	6	5	2	0	0	0	1	5	6	6
Carnivorous beetles	0	1	2	3	5	6	4	2	2	1	0
Moth caterpillars	0	0	0	6	10	12	8	4	0	0	1
Spiders	0	0	0	1	3	4	2	1	0	0	0

a) Make a bar chart for each of the animals in the table.

b) Suggest a reason for the numbers of herbivorous beetles at stations 0–5.

c) Suggest reasons for the numbers of moths at stations 0–5.

d) Suggest reasons for the numbers of carnivorous beetles and spiders at stations 0–5.

e) How do the numbers of animals at stations 6–10 compare with those at stations 0–5?

f) Identify a result that does not fit in with the other results and offer an explanation for it.

1 Use a world map in an atlas to locate your country, then find it on the map on this page.

a) About how long ago did people reach where you live now?

b) How long did it take people to migrate from Africa to where you live now?

c) When people left what is now India, which present-day countries did they pass through to reach:
   i) New Zealand
   ii) South America?

◆ The spread of people around the world
◆ Life in early times
◆ Early machines – watermills and windmills
◆ Steam engines
◆ Generating electricity
◆ Transport
◆ People today
◆ Changes in the environment
◆ Time to save the planet?

Scientists believe that the human species first developed in East Africa about 200 000 years ago, gradually spread north over the next 100 000 years and then moved into the lands to the east and the west.

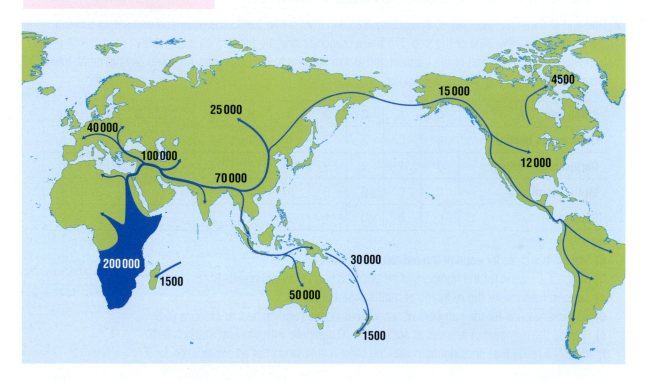

Figure 6.1 The arrows show the paths people took as they migrated across the world. The numbers show how long ago people reached the various parts of the world.

Early times

Over this 200 000 year period, human life has changed for most people. Only a few people today live as the earliest people did – as hunter-gatherers, as described on pages 10 and 11. These people lived in small groups and collected enough food for themselves from their surroundings. When food supplies fell, they moved on, letting the plants and animals breed back to their former numbers, so in the future the people could return and harvest them again. The people also used materials, such as wood for fuel and making temporary buildings, as they moved around.

Figure 6.2 A temporary village of present-day hunter-gatherers in southern Africa

Human activity may influence the environment in a positive or a negative way, or it may not influence it at all and be considered as neutral. We can use these three ways of affecting the environment to give the activity an Environmental Impact Rating (EIR).

2 What is the EIR of hunter-gatherers – positive, negative or neutral? Explain your answer.

3 What is the EIR of the early farmers – positive, negative or neutral? Explain your answer.

4 What is the EIR of setting up towns and cities – positive, negative or neutral? Explain your answer.

About 10 000 years ago, people living in the lands that are now Turkey, Iraq, Iran and parts of Saudi Arabia began farming. Afterwards farming spread to Egypt and North Africa, and into Europe and Asia. Farming involved clearing the land of its natural habitats to make fields for the growing of crops (see Figure 6.3).

Farming provided food for larger numbers of people, who gathered into villages, towns and eventually cities. Farming also allowed some people to work in other occupations, such as pot making, furniture making, cloth making and metalworking, and a trade in goods built up between neighbouring countries, reaching from Ireland in the west to China in the east. During this time, the sources of power for farming, for making goods and for transport were the people themselves and the animals they domesticated, such as cattle, horses and asses.

Animal power was much greater than human power. For example, it could take people ten weeks to prepare an

Figure 6.3 This view across the countryside in England shows how the forests have been cleared to make fields.

area of land for crop growing by using hand tools but only eight days to prepare the same area of land with a plough pulled by oxen.

For most of human history, wood has been used as a fuel to provide heat energy for cooking, for keeping warm and for preparing materials such as pottery and metals. Fuel wood is still used in many places today.

5 When can gathering fuel wood be considered as being environmentally neutral?

Figure 6.4 Carrying home fuel wood in Africa

Early machines

For discussion

What EIR would you give to watermills and windmills? Explain your answer.

About 2200 years ago, the Ancient Greeks invented a machine that used the power of running water to grind up flour. This machine was the watermill. It had a wheel that was turned by the water and a mechanism to transfer this turning motion to stones for grinding up flour. In time, water power was used for a wide range of processes such as sawing wood, weaving and pumping bellows on a forge to provide heat for metal workers. The first windmills were invented between 1500 and 1100 years ago in Persia, in the land that is now Iran. The sails turned round a vertical axis. Later, in Europe, the windmill was developed with a horizontal axis. Its main purpose was grinding grain and pumping water.

Figure 6.5 This early windmill in Greece has 12 small sails to catch the wind. Later windmills such as those in Holland have four large sails.

Steam engines

In 1698, in England, Thomas Savery invented the steam engine. It produced an up-and-down movement and was used for pumping water out of mines, but because of occasional boiler explosions other inventors got to work on improving it. By 1783, James Watt had built a much safer steam engine, which could produce rotary motion like water wheels and windmills. In time, these engines became very powerful and were used to drive all kinds of machines in factories, from paper mills to iron works, as the **Industrial Revolution** got under way. They required large amounts of fuel. This could not be provided by the wood that was available so coal was used instead.

People flocked to the cities to work in the factories as they could earn more money there than by staying in the countryside and working on farms. Many industrial towns grew to the size of cities and the smoke released from the factory chimneys polluted the air above them.

6 What is the EIR of the Industrial Revolution – positive, negative or neutral? Explain your answer.

Figure 6.6 Women working in a cotton mill in Northern England. The mill was powered by a steam engine. As the engine wheel turned, the belts and wheels transferred its power to the machines and made them work.

Generating electricity

In the first half of the 19th century, an English scientist called Michael Faraday found that, if a magnet was made to move quickly past a wire, a current of electricity was generated in the wire. This discovery led to the development of the electrical generator, in which a large magnet is made to spin around inside an even larger coil of wire and produce electricity. Today in many power stations the magnet is attached to a turbine, which is made to spin by driving steam over it.

The water to provide the steam was originally heated by burning coal, and many power stations still use coal today, but other sources are also used. These include oil and gas, which – like coal – are **fossil fuels** and **non-renewable**,

Figure 6.7 Cottam power station in Nottinghamshire, UK, uses coal to generate electricity.

Figure 6.8 Hydroelectric power stations are built in many mountainous regions where a tall dam can be built to store water to power the turbines.

and uranium, a **nuclear fuel**, which is also non-renewable. Some power stations use **renewable** sources of energy. These include water power at hydroelectric power stations and wind turbines at a wind farm.

There are other sources of renewable energy in addition to wind and water power. These include geothermal energy from the hot rocks in the Earth, solar energy from sunlight and energy from burning the unwanted remains of crop plants, such as sugar cane. Energy can also be obtained by growing plants specially for burning, such as oil palms, which are called **biofuels**. Methane gas produced from the wastes stored in **landfill sites** can also be burned to provide energy.

Figure 6.9 The panels on this roof collect energy from sunlight and convert it into electrical energy for use in the house below.

7 What is the Environmental Impact Rating (EIR) of:

a) fossil-fuel power stations

b) nuclear power stations

c) hydroelectric power stations

d) wind farms

e) solar panels

f) methane-burning power stations?

Explain each of your answers.

Figure 6.10 Methane gas produced at a landfill site is being collected using this equipment, for use as a fuel.

Transport

Soon after the steam engine was invented, engineers began to put it to use in transporting people and goods. They developed steam locomotives, in which a steam engine powers wheels to drive the machine along. These locomotives revolutionised transport during the 19th century, because they meant that heavy loads could be carried quickly over large distances. Some are still in use today.

Petrol and diesel engines were developed from the steam engine and have replaced it as the main power source in cars, trucks and buses today. Petrol and diesel are made from oil.

Figure 6.11 This steam locomotive is pulling a train across a bridge in India.

For discussion

How often do you use some kind of engine to make a journey? Keep a journey diary for a week and compare it with the diaries kept by your friends. Identify some ways in which you could reduce the amount of carbon dioxide your journeys produce.

Jet engines were developed in the middle of the 20th century and are now used on most aircraft carrying people and goods around the world. These run on a fuel called kerosene that is also made from oil.

The fuels used in steam engines produce carbon dioxide gas. Petrol, diesel and jet engines also produce carbon dioxide gas, which helps the atmosphere to hold on to heat from the Sun (see 'The greenhouse effect' on page 100).

People today

Many people still live as people did in the past – they use animals to provide power, and fuel wood for cooking and keeping warm. In countries where industry has developed, many people have electricity in their homes and workplaces, they travel in vehicles with petrol and diesel engines and some of the goods they buy, especially fresh food which can turn bad quickly, is flown in from other countries on jet aircraft.

Figure 6.12 Towns and cities around the world consume vast amounts of energy, food and materials to make all kinds of goods.

Figure 6.13 The goods on sale in supermarkets come from all around the world.

8 Use Figures 6.12 and 6.13 to help you to assess the EIR of life in a city. Explain your answer.

Changes in the environment

Changes on land

When people began farming, they destroyed their local habitats to make fields to grow crops and keep livestock. The destruction of the habitats destroyed the organisms and the food chains that bound them together. Today farmers need to produce huge amounts of food for the world's human population, so pests that attack crops have to be killed. Chemicals called **pesticides** are used and some can be harmful to other animals.

Figure 6.14 A helicopter spraying trees in an orchard with pesticides

The pesticide slowly builds up in the body of the pest as it feeds. If it is eaten by a predator, such as an insectivorous bird, the pesticide passes further up the food chain. If this bird is then eaten by a tertiary consumer, such as a sparrowhawk, the pesticide is passed on and stored in the hawk's body. After eating many of its prey, the hawk can receive a fatal dose or the poison may cause it to lay weak-shelled eggs, which fail to survive. Not all pesticides are as harmful as this and many governments have rules for farmers to follow to protect the environment.

When factories and trade began to grow and large amounts of fuel for steam engines were needed, coal mines were dug or the ground surface was ripped off to make an open cast mine, destroying the habitats and their food chains. As factories began to produce more and more goods, they needed large amounts of materials such as metals. Rocks containing the metals, called **ores**, were dug out of the ground in open cast mines in many parts of the world, destroying even more habitats and food chains.

9 Draw ten outlines of an insect and put a dot in each one to represent a dose of pesticide. Now draw an arrow leading from the insects to five birds, and draw two dots in each to show how the concentration of poison builds up. Complete the food chain by drawing a single hawk with its dose of poison.

10 Identify four ways in which people have destroyed food chains.

Figure 6.15 Open cast mining in a rainforest in Papua New Guinea

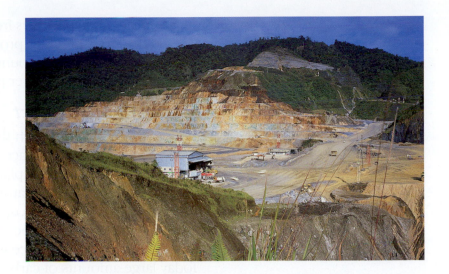

Changes in the sea

Oil can be taken out of the ground without too much environmental damage but if there is a break in an oil pipe at sea, or if an oil tanker runs aground while transporting it, a spillage can cause a great deal of damage for many years.

Figure 6.16 A sea bird contaminated by oil from a spillage

11 How are the food chains in a beach habitat affected when oil covers a beach?

Changes in the atmosphere

The first atmosphere on the Earth was produced by the gases escaping from erupting volcanoes about five billion years ago. These gases were water vapour, carbon dioxide and nitrogen. Three billion years ago the first plants developed and – as they made food from carbon dioxide and water, using the energy in sunlight – they produced oxygen, which entered the atmosphere too. About 25 km above the

Earth's surface, ultraviolet rays from the Sun reacted with oxygen in the atmosphere to produce a layer of ozone. This screened out a large amount of harmful rays from the Sun and made the world a safer place for living things.

The greenhouse effect

When rays from the Sun reach the atmosphere, they pass through it to the Earth's surface and some of the heat energy they carry warms the planet. Some of this heat is radiated back through the atmosphere to space. As it travels back through the atmosphere some of it is absorbed by carbon dioxide. This trapped heat also warms the planet and has helped make it a place where life can thrive.

Today large amounts of carbon dioxide are entering the atmosphere due to human activities, such as burning fossil fuels in power stations, and using huge numbers of vehicles that release carbon dioxide in their exhausts. There is evidence that the climate is beginning to change and that the planet will become warmer. There is also evidence that the Earth has warmed and cooled a little in the past, naturally. However, it could be that this time the extra carbon dioxide produced by human activity is adding to the heating effect and this may produce great changes to habitats around the world.

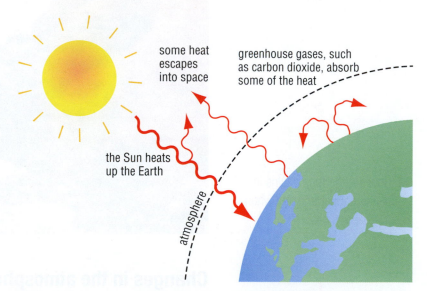

some heat escapes into space

greenhouse gases, such as carbon dioxide, absorb some of the heat

the Sun heats up the Earth

atmosphere

Figure 6.17 The path of heat between the Sun and the Earth

Acid rain

There are some gases occurring naturally in the atmosphere that dissolve in the water droplets in clouds and make the rain slightly **acid**. Some of these gases contain sulfur and are produced by erupting volcanoes and algae that live in the sea. Some other acid gases, containing nitrogen, are produced by lightning reacting with nitrogen in the atmosphere. They are also released when fossil fuels are burnt at power stations and factories and in petrol and diesel engines. In some places, the acid gases from power stations are blown by prevailing winds in one direction – say, to the west – for a long time. When this happens, rain produced to the west of the power stations is unusually acid and this causes soil damage, which kills plants.

12 What happens to the food chains in a forest when the forest is affected by acid rain? Explain your answer.

Figure 6.18 A forest of spruce trees damaged by acid rain in Europe. Acid rain also occurs in Taiwan, China and part of the United States and Canada.

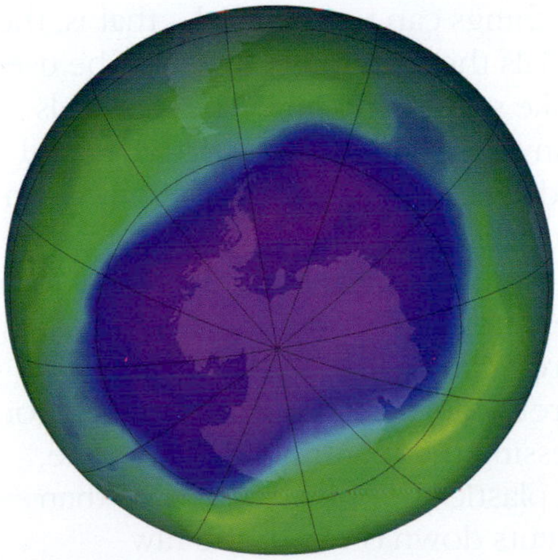

Figure 6.19 The hole in the ozone layer is shown by the purple colours. The hole varies in size throughout the year and when this image was made it covered almost the whole of Antarctica.

The ozone layer

In the 1920s, gases called chlorofluorocarbons (CFCs) were made for the first time. They were used for keeping things cool in fridges and air conditioning, and because they squashed easily they were used in aerosol sprays. The CFCs entered the atmosphere. In the 1980s, scientists noticed that holes were developing in the ozone layer around the North Pole and South Pole. When they investigated this, the scientists found that CFCs destroy ozone in the atmosphere and since then the governments of 196 countries have agreed to reduce the use of these chemicals.

Time to save the planet?

It is clear that human activity has had a major influence on the environment, resulting in habitat destruction and an increasing number of endangered species. It could also be making the planet a less suitable place for human life. Many people realise this and have begun to develop activities to reduce the environmental damage that has been done in the past and is still happening today. Below are some examples of such activities.

Cutting energy use

To save energy, both at home and in the workplace, we should switch off lights and electrical equipment that is not in use and not leave televisions and computers on 'standby'. Low energy light bulbs can be fitted in the home, and washing machines could be used only when there is a full load to wash. Clothes should be dried outside whenever possible, instead of using a tumble dryer. When the temperature begins to fall, we can save energy by putting on warm clothes, rather than having the central heating thermostat turned up high.

Recycling and re-using

Many of the things we use are thrown away eventually. A newspaper is thrown away after a day, a plastic lemonade bottle after perhaps three days and some items of clothing after about a year. All of these things can be recycled – that is, the materials they are made from can be used to make something new. The materials in many products, such as batteries and cartridges for computer printer inks, can be recycled too.

Re-using products in their original form is even more environmentally friendly than recycling the materials from which they're made, as no energy is needed for processing the materials. For example, using plastic shopping bags more than once cuts down on both the raw materials needed to make new bags and the energy used in recycling plastic from

13 Using the information in this section, design a poster featuring the Earth and the ways its environment has been damaged. The purpose of the poster is to motivate people to help save the planet so you might want to show the Earth with a sad-looking or horror-struck face, for example.

14 Are you wasting energy at home? Explain your answer. What could you do to help save the planet?

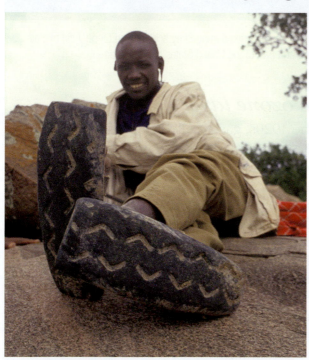

Figure 6.20 This man is wearing shoes made from old car tyres.

15 One way to recycle some materials would be to use them as packaging and wrapping for presents. Suggest some materials that you could use (other than old wrapping paper) and think about how you could make them into an unusual wrapping for the next presents you give.

16 How do you think addressing these five problems will affect the food chains in habitats of a damaged environment?

17 How many national parks are in your country? Use the internet to find out. Which one is nearest to where you live?

18 How will the management of a national park affect the food chains in the park's habitats?

For discussion

Look back at the section on food chains on page 67 and find out about some food chains that exist among the plants and animals in your country. Discuss how you think the activities of people affect the food chains in a positive and a negative way.

old bags. Using large cloth bags is better still, because they are made from a renewable material rather than plastic (which comes from oil), and they last much longer than plastic bags.

The work of governments and companies

While everyone can alter their life style to help the planet, some problems can only be tackled by large organisations such as companies and governments. Large-scale projects to address such problems include:

- setting up recycling schemes
- developing alternative ways of generating electricity to reduce the effects of burning fossil fuels
- reducing the use of CFCs
- reducing the use of harmful pesticides in farming
- developing devices that reduce pollution, such as catalytic converters fitted to the exhaust pipes of petrol and diesel engines to remove harmful gases.

These approaches are aimed at reducing the impact of human activities on the damaged environment, allowing it to recover over time. Many countries have also set aside protected areas where the aim is to prevent environmental damage occurring in the first place. These areas are known as national parks, nature reserves or wildlife sanctuaries. The species of plants and animals living in them are monitored using a variety of techniques, including those used in Chapter 5, and damage to the environment is prevented by enforcing a set of special regulations and laws.

Figure 6.21 This Bengal tiger cub is in the Bandhavgarh National Park in India.

◆ SUMMARY ◆

◆ People spread out from Africa to populate the Earth (*see page 90*).

◆ The use of animals and the plough improved farming and impacted on the environment (*see page 91*).

◆ The development of towns and cities had an impact on the environment (*see page 91*).

◆ Early machines such as the watermill and windmill used renewable energy sources (*see page 93*).

◆ Steam engines and the Industrial Revolution changed the way of life for many people and also had an impact on the environment (*see page 93*).

◆ Fossil fuels and nuclear fuels for power stations are non-renewable energy sources (*see page 94*).

◆ Hydroelectric power stations and wind farms use a renewable energy source (*see page 95*).

◆ Sunlight is a renewable energy source (*see page 95*).

◆ Engines used in transport cause pollution (*see page 96*).

◆ Some pesticides can also kill animals other than pests (*see page 98*).

◆ Mining destroys habitats (*see page 98*).

◆ Oil spills damage sea life (*see page 99*).

◆ Some human activities lead to pollution of the atmosphere (*see page 99*).

◆ People can help to improve the environment by altering their life styles at home and in the workplace (*see page 102*).

◆ Governments and large companies can improve the environment by setting up large-scale projects and enforcing laws and regulations (*see page 103*).

End of chapter questions

1 Imagine that you live in a country with supplies of coal, oil and uranium, and mountains with rivers running down deep valleys. There are also windswept plains and long, sunny summers. There is just one town in your country. How would you plan to provide the people with electricity with the minimum damage to the environment?

2 In order to find out about the environmental impact of human activities on an area, records must be kept of its natural habitats for a number of years.
   a) If you have a 'wild area' in your school grounds, or if there is a habitat such as a wood or a group of bushes in your neighbourhood, how would you set up an investigation to check for signs of change in the populations of the living organisms in this area over the years?
   b) What results would you predict if there was no environmental impact of human activities?
   c) What results would you predict if human activities were having an environmental impact?

7 Classification and variation

Classifying living things

Living things are put into groups so that they can be studied more easily. Grouping in this way is called **classification**. The largest groups in the classification system for living things are called **kingdoms**. In Chapter 4, the kingdoms of the microorganisms were described and in this chapter we will look at the plant and animal kingdoms. Each kingdom contains a large number of living things that all have a few major features in common. Table 7.1 shows the features that are used to place living things in either the plant or the animal kingdom.

Table 7.1 The features of living things in the plant and animal kingdoms

Plant kingdom	Animal kingdom
make their own food from air, water, sunlight and chemicals in the soil	cannot make their own food – eat plants and animals
body contains cellulose for support	body does not contain cellulose
have the green pigment chlorophyll	do not have chlorophyll
stay in one place	move about

Dividing up the animal kingdom

The way that the animal kingdom is divided up into subgroups is described on the following pages. A similar way of subgrouping is used to divide up the plant kingdom.

The subgroups of the animal kingdom can be put into two groups called the invertebrates and the vertebrates. **Invertebrates** do not have an inside skeleton of cartilage or bone. **Vertebrates** do have an inside skeleton of cartilage or bone. These two groups can be divided further.

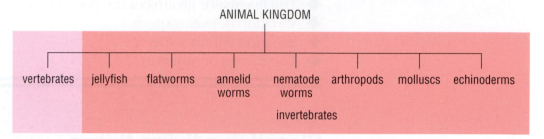

Figure 7.1 The major subgroups of the animal kingdom

Groups of invertebrates

1 How is an earthworm different from a wasp? How is it similar to a wasp?

The main subgroups in the invertebrate group are the jellyfish, flatworms, annelid worms, nematode worms, arthropods, molluscs and echinoderms.

jellyfish (compass jellyfish)   flatworm (pond flatworm)   annelid worm (earthworm)   nematode worm (roundworm)

arthropod (wasp)   mollusc (snail)   echinoderm (starfish)

Figure 7.2 Examples of invertebrates

Annelids

Annelids have long, thin, soft bodies divided into segments or rings.

Figure 7.3 An earthworm is an annelid.

Nematodes

Nematodes have thin, cylindrical bodies that are not divided into segments.

Figure 7.4 A highly magnified view of a nematode worm in the soil. The picture has been coloured to make the worm easier to see.

Arthropods

The name arthropod means 'jointed leg'. All arthropods have a skeleton on the outside of their body, and jointed legs.

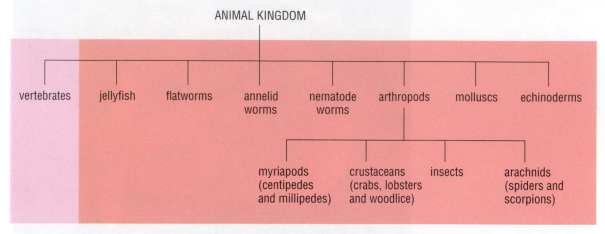

ANIMAL KINGDOM

vertebrates · jellyfish · flatworms · annelid worms · nematode worms · arthropods · molluscs · echinoderms

myriapods (centipedes and millipedes) · crustaceans (crabs, lobsters and woodlice) · insects · arachnids (spiders and scorpions)

Figure 7.5 The classes of arthropods in the animal kingdom

There are four major subgroups of arthropods, called **classes**.

- Myriapods have one pair of antennae (*singular*: **antenna**) and long cylindrical or flat bodies with many legs. The centipedes and millipedes are in this group (Figure 7.6).
- Crustaceans, like the lobster and woodlouse, have two pairs of antennae (Figure 7.7).
- Arachnids do not have antennae or wings but have four pairs of legs. Spiders, mites, ticks and scorpions belong to this group (Figure 7.8).
- Insects have one pair of antennae, three pairs of legs and up to two pairs of wings (Figure 7.9).

Figure 7.6 A giant millipede on the floor of a rainforest. These myriapods are found in tropical regions of Africa, America and Asia.

Figure 7.7 This lobster, walking along the seabed, is an example of a crustacean.

Figure 7.8 A black widow spider – one of the most venomous arachnids in the world. Black widow spiders are found in America, countries around the Mediterranean Sea, Asia, Australia and New Zealand.

Figure 7.9 The Monarch butterfly is a particularly beautiful example of an insect species. It is found all over the world in tropical and subtropical areas.

Molluscs

The mollusc group gets its name from the Latin word *mollis*, which means 'soft'. This refers to the soft bodies of the animals. Most molluscs have a shell to protect their body. A snail has a coiled shell from which it pushes out its head and fleshy 'foot' when it wishes to move and feed. The slug also has a small shell, but it is under the saddle-like structure (called the mantle) on its back, and does not provide protection.

Figure 7.10 An octopus on a coral reef. An octopus is a mollusc with tentacles.

Vertebrates – an example of how living things are classified

In the vertebrate group are five major classes. They are the fish, amphibians, reptiles, birds and mammals.

Table 7.2 The features of five of the vertebrate classes

Bony fish	Amphibians	Reptiles	Birds	Mammals
Scales, fins Eggs laid in water	Smooth skin Eggs laid in water	Scales Soft-shelled eggs laid on land	Feathers Hard-shelled eggs	Hair Suckle young with milk

2 How is a goldfish different from a frog, and how is it similar?

Each class is divided up into smaller groups called **orders**. The members of each order have so many features in common that they look alike and are easy to group. There are 19 orders of mammals. Four examples are shown in Table 7.3.

Table 7.3 Four orders of mammals

Insectivores	Bats	Rodents	Whales
Small body Long snout	Small body Wings	Chisel-like teeth for gnawing	Flippers Tail with fins

3 The insectivore in Table 7.3 is a shrew and the rodent is a mouse. How are they different and how are they similar to each other?

An order is made up of smaller groups called **families**. The members of the different families look similar but there are differences. This can be seen by looking quite closely, as shown in the four families of whales (Table 7.4).

Table 7.4 Four families of whales.

Beaked whale	Sperm whale	Dolphin	White whale
Few teeth Small flippers	Large head Rounded back fin	Sickle-shaped flippers and back fin	No back fin Blunt snout

4 How is a beaked whale different from a sperm whale?
5 How is a beaked whale similar to a sperm whale?

The members of a family have differences between them and are split up into smaller groups called genera (*singular*: **genus**). The differences between members of each genus are found by looking very closely. For example, if you look at dolphins A, B and C in Figure 7.11, you will see that A seems to have more features in common with B than with C. Because of this, dolphins A and B are placed in one genus and C is placed in a separate genus.

A Dusky dolphins live in coastal waters off New Zealand, Africa and South America.

B The white-sided dolphin lives in the open water of the northern Pacific Ocean and around the coasts of Japan and North America.

Figure 7.11 Members of the dolphin family

C Bottle-nosed dolphins live in all the oceans and seas of the tropical and temperate regions of the world.

Because dolphins A and B have small differences between them, they are placed in separate groups called **species**. A species is a group of animals that have a very large number of similarities, and the males and females breed together to produce offspring that can also breed. The males and females of different species do not normally breed together, but if they do they produce offspring that are usually sterile (cannot breed). For example, if a male donkey and a female horse mate, the offspring they produce is a sterile mule.

6 How are dolphins A and B different from C?

7 How are dolphins A and B different from each other?

International names of living things

About 500 years ago, the Age of Discovery began in Europe when explorers from Spain, Portugal, France, England and the Netherlands began to sail the world's oceans to discover new lands and new people with whom to trade. Over the next 300 years or so, explorers took back plants and animals from the places they visited.

There was also an increasing interest in science at this time. As scientists studied the unfamiliar living things that explorers brought home, they tried to sort them into groups to make their work easier.

Little progress was made until Carl Linnaeus (1707–1778) had an idea. He lived in Sweden but travelled all over western Europe collecting and studying plants. He made close observations of his specimens and worked out a way of describing how one living thing was different from another. He began by putting very similar living things in the same group. This group is called the genus or general group. He then looked at all the specimens in the genus and sorted them out into smaller groups in which all the individuals were alike. This group is called the species or special group.

Linnaeus then decided to give each group, the genus and species, a name. Each name should describe some feature of the specimen. He further decided that the two names became the scientific name of the living thing. At the time, scientists in every country learnt and used Latin and Greek so he used words from these two languages so that every scientist could understand them.

These words are still used today. For example, the genus and species names of the African clawed toad are *Xenopus laevis. Xenopus* is made from two Greek words – *xenos* meaning 'strange' and *pous* meaning 'foot'. The words refer to the toad's webbed hind foot, each toe of which is capped with a dark, sharp claw. The word *laevis* is Latin for 'smooth' and refers to the toad's smooth skin.

Figure A Carl Linnaeus, dressed for a journey into the Scandinavian countryside to look for plants

1 Why was there a need for grouping living things as explorers brought them back from their travels?
2 How old was Carl Linnaeus when he died?
3 Why were Latin and Greek used to name living things?

4 Which scientific activities led Linnaeus to think of his idea for classification?

For discussion

Why were the common names or local names not used in the naming of plants and animals by scientists?

Figure B *Xenopus laevis* lives in wetlands in South Africa.

The plant kingdom

The plant kingdom is divided into five groups – the algae, liverworts and mosses, ferns, conifers and flowering plants.

Figure 7.12 Seaweeds have a root-like holdfast to grip rocks and stop them being washed away.

Algae

Algae are very simple plants. They do not have roots, stems or leaves but they do contain the green pigment or colouring called **chlorophyll**, which is a characteristic of all plants. Some of the largest species – the seaweeds – also have red pigments and brown pigments.

Most species of algae are so small they can only be seen with a microscope and some scientists put them in the Protoctista kingdom (see Chapter 4). When they occur in large numbers, they can be clearly seen and they turn pond water green, form bright green patches on tree trunks and create slime on rocks in streams. Algae form a large part of the plankton that is found in the upper parts of seas and oceans.

8 Why do seaweeds not grow on sandy beaches, although they may still be found there?

Liverworts and mosses

Liverworts are small plants that do not have true roots, stems or leaves. They grow in damp places near streams and ponds.

Mosses have stems and leaves but they do not have proper roots. Moss plants are usually found growing together, in many different habitats from dry walls to damp soil.

Both liverworts and mosses reproduce by producing spores. They make the spores in a capsule that is raised into the air. When the capsule opens the spores are carried away by air currents.

9 How is a liverwort different from a moss?

Figure 7.13 A liverwort (left) and a moss plant (right)

Ferns

Ferns have true roots and stems and reproduce by making spores. These are made in structures called sporangia on the underside of large feather-like leaves called fronds. When the sporangia open, the spores are released into the air.

10 In what way are ferns similar to liverworts and mosses?

Figure 7.14 Ferns grow in many damp, shady places around the world.

Conifers

A conifer has roots, a woody stem and needle-like leaves. Most conifers lose and replace their leaves all year round, so they are called evergreen. Almost all conifers reproduce by making seeds that develop in cones. When the seeds are ready to be dispersed the cones open and the seeds fall out. Each seed has a wing that prevents the seed falling quickly and allows it to be blown away by the wind.

Flowering plants

A flowering plant has a root, stem and leaves. In some plants, the stem is woody. All of these plants reproduce by flowering and making seeds.

Figure 7.15 The needle-like leaves of conifers mean they are better able to survive in frozen conditions than broad-leaved trees.

11 In what ways is a conifer different from a moss?
12 In what way is the reproduction of conifers, mosses, liverworts and ferns similar?

Figure 7.16 Grass, bluebells and these trees are all flowering plants. These are growing in a woodland in Europe.

Variation

Variation between species

Many living things have certain features in common. For example, a cat, a monkey and a rabbit all have ears and a tail. However, these features vary from one kind of animal to the next. In the species shown in Figure 7.17, the external parts of the ears of the rabbit are longer than the ears of the cat. The external parts of the monkey's ears are on the side of its head while the other two animals have them on the top. The cat and the monkey have long tails but the monkey's tail is prehensile, which means the monkey can wrap it around a branch for support while it hangs from a tree to collect fruit. (Only monkeys that come from South America have prehensile tails.) A rabbit's tail is much shorter than the cat's tail and the monkey's tail.

These **variations** in features are used to separate living things into groups and form a classification system (see page 105) that is used worldwide.

Figure 7.17 A cat, a rabbit and a Costa Rican spider monkey have features in common, but they show variations too.

Variation within a species

The individuals in a species are not identical. Each one differs from all the others in many small ways. For example, one person may have dark hair, blue eyes and ears with lobes while another person may have fair hair, brown eyes and ears without lobes. Another person may have different combinations of these features. There are two kinds of variation that occur in a species. They are continuous variation and discontinuous variation.

Continuous variation

A feature that shows **continuous variation** may vary in only a small amount from one individual to the next, but when the variations of a number of individuals are compared they form a wide range. Examples include the range of values seen in heights or body masses.

Discontinuous variation

A feature that shows **discontinuous variation** shows a small number of distinct conditions, such as being male or female, and having ear lobes or no ear lobes. There is not a range of values between the two, as there is between a short person and a tall person, for example. However, there are very few examples of discontinuous variation in humans.

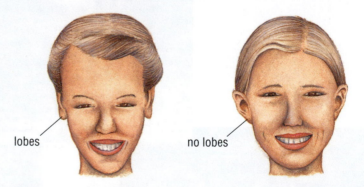

Figure 7.18 Ears with and without ear lobes

The causes of variation

Some members of the family in Figure 7.19 have similar features. They are found in different generations, which suggests that the features could be inherited. Some variations may also be due to the environment.

Figure 7.19 Members of a family

For discussion

Look at this photograph of a family. How are the two boys like their mother? How is their father on the left like their grandparents on the right?

Variation and the environment

The **environment** can affect the features of a living organism. For example, if some seedlings of a plant are grown in the dark and some in the light, they will have different features. Those grown in the dark will be tall, spindly and have yellow leaves, while those grown in the light will have shorter, firmer stems with larger leaves that are green.

Lack of **minerals** in the soil can affect the colour and structure of the leaves. For example, the presence of lime in the soil affects the colour of hydrangea flowers. If the soil contains lime, the flowers are pink. If the soil is lime-free, the flowers are blue (see page 150).

The colour of flamingos is also affected by the environment. The flamingo feeds on shrimps that possess a pink pigment. The pigment passes into the feathers and makes the flamingo pink.

13 How else could the environment affect the development of an organism? Give another example for a plant and for an animal.

Figure 7.20 When pink flamingos do not eat enough shrimps their feathers become white. Flamingos live in aquatic habitats in Africa, Asia, the Caribbean, Europe and South America.

The food an animal eats often affects variation within the species. If the environment does not provide enough food, adult animals become thin and have a smaller body mass. Young animals grow slowly so they look smaller than other members of the species who have enough food and are the same age.

Usually animals eat only enough to keep them healthy and do not become too fat. However, animals that hibernate, like bats, or sleep for long periods of time through the winter, like bears, eat large amounts of food in the autumn and increase their body mass. The fat that they store provides them with enough energy to keep them alive while they are sleeping and not able to feed. When they emerge in the spring their body mass is greatly reduced.

Humans who eat too much food increase their body mass and this may threaten their health. In some countries, the variation in body mass in the population is due to some people eating far too much.

◆ SUMMARY ◆

◆ Living things are classified by putting them into groups (*see page 105*).

◆ There are five kingdoms of living things. They are the animals, plants, fungi, Monera and Protoctista (*see page 105*).

◆ The animal kingdom can be put into two groups called the invertebrates and the vertebrates (*see page 106*).

◆ The main subgroups of the invertebrates are the jellyfish, flatworms, annelid worms, nematode worms, echinoderms, molluscs and arthropods (*see page 106*).

◆ There are four major classes of arthropods. They are the myriapods, crustaceans, arachnids and insects (*see page 108*).

◆ There are five classes of vertebrates. They are the fish, amphibians, reptiles, birds and mammals (*see page 110*).

◆ When a living thing is classified it is assigned to a class, order, family, genus and species (*see page 110*).

◆ The plant kingdom can be put into five groups. They are the algae, liverworts and mosses, ferns, conifers and flowering plants (*see page 113*).

◆ There is variation between species (*see page 115*).

◆ There are two kinds of variation within a species. These are continuous and discontinuous variation (*see page 116*).

◆ The environment can affect the variation in a species (*see page 117*).

End of chapter questions

1 What kind of living organism is each of the following?
a) does not have a backbone but has five arms
b) has a backbone and wings
c) does not have a backbone but has wings
d) has scales and lays eggs in water
e) has scales and lays eggs on land
f) has a backbone and hair

2 Imagine that you have landed on a distant planet. When you climb out of your spacecraft you find some small eight-legged, six-eyed animals leaping about. You call them 'hoppies', and gather information about 20 of them in a table.
a) Arrange the hoppies into five size groups based on their mass.
b) Display the numbers in the groups in a bar chart.
c) Arrange the hoppies into groups based on colour.
d) Display the numbers in the groups in a bar chart.
e) Which feature shows continuous variation?
f) Which feature shows discontinuous variation?
g) Is there any relationship between the mass of the hoppies and their colours? Describe what you find.
h) Speculate on a reason for your findings.

Hoppy	Mass/g	Colour
1	200	green
2	349	green
3	210	green
4	615	blue
5	430	yellow
6	570	red
7	402	yellow
8	429	yellow
9	317	green
10	520	red
11	460	yellow
12	403	yellow
13	330	green
14	489	yellow
15	502	red
16	630	blue
17	410	yellow
18	380	green
19	550	red
20	445	yellow

CHEMISTRY

◆ There are three states of matter – solid, liquid and gas.
◆ Matter is made up of particles, which are arranged differently in solids, liquids and gases.
◆ Matter can change from one state to another.
◆ Changes in state can be explained by the particle theory.

What is everything made from? Scientists have asked this question for thousands of years and gradually they came up with an answer. Everything is made of **matter**.
There are three forms, or states, of matter – the solid state, the liquid state and the gaseous state. This means that everything is made up from materials that are a solid, a liquid or a gas.

Matter everywhere

To help you think of the world in terms of the three states of matter, think about going on a school hike. As you walk along, you move across the solid Earth. Your body pushes through a mixture of gases we call the air. If it rains as you walk along, droplets of liquid fall from the sky.

For discussion

Select one state of matter and imagine that it has been removed from the world. List things that could not exist if it was absent.

Do the same for the other two states of matter. Would it be possible to live in any of the three imaginary worlds?

Figure 8.1 The three states of matter on a school hike

Not only do you move through a world made from the three states of matter, you are made from them too. You have solid bones that are moved by solid muscles, as liquids flow through your arteries, veins and intestines. When you breathe in, air – that mixture of gases – fills your windpipe and your lungs.

Comparing the states of matter

You can tell one state of matter from another by examining their properties.

- A **solid** has a definite mass, a definite shape and its **volume** does not change. It does not flow and it is hard to compress (squash) it.
- A **liquid** has a definite mass and its volume does not change. It is hard to compress but it flows easily. The shape of the liquid varies and depends on the shape of the container holding it.
- A **gas** has a definite mass but its volume can vary and it takes up the shape of the container holding it. It flows easily and it is easy to compress.

1 Make a table of the properties of the three states of matter.
2 How are all three states of matter:
   a) similar
   b) different?
3 Identify the three states of matter that make up a glass of fizzy drink.

Figure 8.2 This picture shows the three states of matter – solid (the ice), liquid (the drink) and gas (the bubbles).

The first ideas about matter

The earliest people used the materials they could find around them, such as wood, stone, antlers and skin. When people learned to make fire, they began to change one material into another. First they learned how to cook food, then how to bake clay and make pottery and bricks. Eventually they learned how to heat some rocks in charcoal fires so strongly that a chemical reaction took place in which a metal was produced.

By 600 BCE, philosophers in the Greek civilisation were thinking about what different things were made of. They were puzzled by the way one substance could be changed into another. They asked the question, 'If a rock can be turned into metal, what really is the rock? Is the rock a kind of metal or is the metal a kind of rock?' They then thought that if one substance could change into another, perhaps it could go on changing into other substances. They did not carry out experiments to test their observations and ideas but tried to explain them with more ideas.

A Greek philosopher called Thales (624–546 BCE) believed that all substances were made from different forms of one single substance. He called this substance an element. He observed how water changed from solid to liquid to gas and how plants and animals needed water to stay alive. From these observations he concluded that everything was made from different forms of water.

Other philosophers did not agree with Thales. Some believed that everything was made from air. They believed that air reached up from the ground and filled the whole of space. They thought that air could be squashed to make liquids and solids. Some philosophers suggested that fire was the basic element because it was always changing and it was this element in everything that made things change.

Eventually it was agreed that there were four elements from which all matter was made. The elements were air, water, earth and fire. Each element was given properties, and the way that the elements and their properties were related to each other is shown in Figure A.

1 Why was the discovery of how to make fire important in making people think about the structure of materials?
2 Why were the Greeks' conclusions about matter not scientific?
3 In what ways do you think Thales saw water change?
4 If a substance was cold and dry, what element did the Greeks think it was made of?
5 What properties would a material have to show for the Greeks to decide that it contained fire?
6 Which elements do you think the Greeks thought were in:
   a) wood
   b) oil
   c) metal?
   Explain your answers.
7 How do you think the Greeks may have explained the changes they saw when a candle burned?

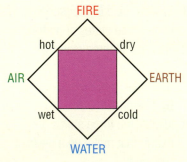

Figure A The Greek elements

Figure B The four elements – air, water, earth and fire – can be easily identified in our surroundings.

The Greeks' ideas of the elements were used for 2000 years to explain the structure of materials and the way they change.

One Greek philosopher called Democritus (around 460–370 BCE) had another idea. He thought about what everything was made of and wondered what would happen if you took a solid substance and divided it in two, then divided each piece in two, and then carried on dividing. He believed that eventually a tiny piece would be produced that was too small to be divided any more. He called this tiny piece of matter an atom. The word atom means 'indivisible'.

The idea about atoms passed on to other scientists living later, and they made investigations that led to a scientific theory called the particle theory. This in turn led scientists to develop apparatus that allows us to see the arrangement of atoms in some materials.

Figure C Democritus was the first person to suggest that matter is made up of atoms.

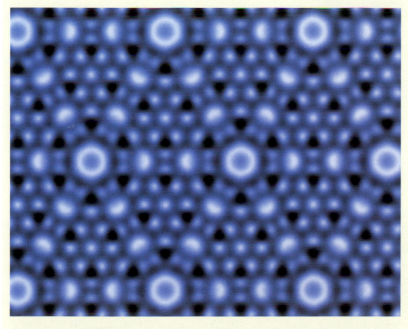

Figure D The arrangement of individual atoms on the surface of a silicon crystal can be a seen with a very powerful microscope called an electron microscope.

8 Why do you think the Greeks' idea of elements was used for such a long time?

9 Tear up a piece of paper as Democritus suggested. How small can you make the paper particles?

10 Look back in more detail at the scientific enquiry section on page 6 and find two activities of the Greeks that are parts of scientific enquiries we use today. What are they?

The particle theory of matter

The **particle theory** states that matter is made from particles. The particles are so tiny that they cannot be seen with the naked eye. Different substances are made from different particles and the particles have different sizes. The particles are **atoms** and **molecules**.

Particles in the three states of matter

The particles in solids

In solids, strong forces hold the particles together in a three-dimensional structure. In many solids, the particles form an orderly arrangement called a lattice. The particles in all solids move a little. They do not change position but vibrate to and fro about one position.

The particles in liquids

In liquids, the forces that hold the particles together are weaker than in solids. The particles in a liquid can change position by sliding over each other.

The particles in gases

In gases, the forces of attraction between the particles are very small and the particles can move away from each other and travel in all directions. When they hit each other or the surface of their container they bounce and change direction.

4 According to the particle theory, why do liquids flow but solids do not?

5 How is the movement of particles in gases different from the movement of particles in liquids?

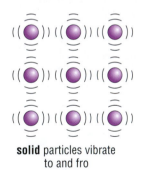

solid particles vibrate to and fro

liquid particles have some freedom and can move around each other

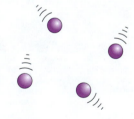

gas particles move freely and at high speed

Figure 8.3 Arrangement of particles in a solid, a liquid and a gas

When matter changes state

The state of matter of a substance can be changed. This kind of change is called a physical change. It is a reversible reaction. This means that the reaction can go forwards and also backwards, as we shall see. A physical change of state can be brought about by heating or by cooling.

Melting

If a solid is heated enough, it loses its shape and starts to flow. This change is called **melting** and the solid turns into a liquid. The temperature at which melting takes place is called the melting point. This can be found by heating a solid and recording its temperature. When the temperature remains constant or steady, the melting point of the solid has been reached. Melting occurs in many substances around us. For example, butter or ghee melts in a pan during cooking and chocolate can melt in your pocket.

Figure 8.4 Ghee being melted in Indian cookery

The particle theory can be used to explain melting in the following way. When a solid is heated, the heat provides the particles with more energy. The energy makes the particles vibrate more strongly and push each other a little further apart – the solid expands. If the solid is heated further, the energy makes the particles vibrate so strongly that they slide over each other and become a liquid. During the time from when the solid starts to melt until it has completely turned into a liquid, its temperature does not rise. All the heat energy is used to separate the particles so that they can flow over one another.

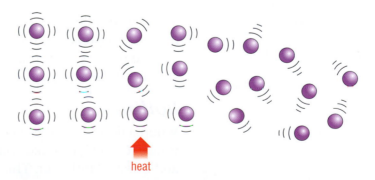

heat

Figure 8.5 The particle arrangement in a solid (left) changes as the heat turns it into a liquid (right).

---

6 What is the heat source that causes the melting of:
a) ghee
b) chocolate in your pocket?

7 The table shows how the temperature of a solid changed as it was heated.

Time/ mins	Temperature/ °C
0	0
1	10
2	20
3	30
4	40
5	50
6	55
7	57
8	59
9	60
10	60

a) Plot a graph for this data.
b) Was the melting point reached? How can you tell?

Figure 8.6 The water flowing down these rocks has frozen to form ice.

8 Why does the wax not freeze at the top of a candle?

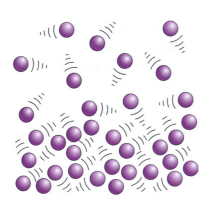

Figure 8.7 Particles evaporating from a liquid surface

Freezing

Freezing is the reverse of melting. It is the changing of a liquid into a solid. The temperature at which this takes place is called the freezing point. It is the same temperature as the melting point of the substance. When we think of freezing we usually think of the freezing of liquid water into solid ice as shown in Figure 8.6.

Other substances can freeze into solids at temperatures much higher or lower than this. For example, when molten wax runs down the side of a candle it freezes and becomes a solid before it reaches the bottom.

The particle theory explains freezing in the following way. If a liquid is cooled sufficiently the particles lose so much energy that they can no longer slide over each other. The only movement possible is the vibration to and fro about one position in the lattice. The liquid has become a solid.

Evaporation

A solid turns into a liquid at one definite temperature (the melting point) but a liquid turns into a gas over a range of temperatures. For example, a drop of water can turn into a gas (known as water vapour) at room temperature of about 20 °C while outside a puddle dries up in the warmth from the Sun. The process by which a liquid changes into a gas over a range of temperatures is called **evaporation**. The gas escapes from the surface of the liquid.

If the temperature of the liquid is higher, it evaporates faster. If the air above the liquid does not already have a lot of vapour in it, the liquid evaporates faster. For these reasons, the speed at which water evaporates into the air is measured as part of a weather survey in many parts of the world, such as North America, Europe and India.

The particle theory can be used to explain evaporation in the following way. The particles in a liquid have different amounts of energy. The particles with the most energy move the fastest. High-energy liquid particles near the surface move so fast that they can break through the surface and escape into the air, forming a gas.

Boiling

When a liquid reaches a certain temperature, it forms a gas inside it. The gas makes bubbles, which rise to the surface and burst into the air. This process is called **boiling**. The temperature at which it takes place is called the boiling point. If the boiling liquid is heated more strongly, its temperature does not rise but it boils more quickly.

Reading from the graph in Figure 8.8, answer the following.

9 What was the temperature of the liquid:

  a) at the start of the experiment

  b) after 2 minutes?

10 At what time was the temperature:

  a) 30 °C

  b) 70 °C?

11 What is the boiling point of the liquid?

12 How did the rate at which the temperature increased change as it reached the boiling point?

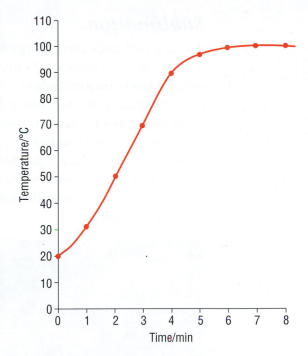

Figure 8.8 Graph to show the boiling point of a liquid

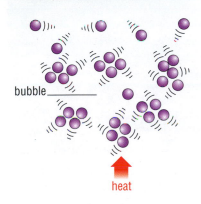

Figure 8.9 A bubble in a boiling liquid

The boiling point can be found by heating a liquid and recording its temperature. When the temperature stops rising and remains constant, the boiling point of the liquid has been reached (Figure 8.8).

The particle theory can be used to explain boiling in the following way. When a liquid is heated, all the particles receive more energy and move more quickly. The fastest-moving particles escape from the liquid surface or collect in the liquid to form bubbles. The bubbles rise to the surface and burst open into the air. The fast-moving particles released from the liquid form a gas.

Condensation

If a gas is cooled down far enough, it turns into a liquid by a process of **condensation**. This process is the reverse of evaporating and boiling.

When water vapour that has evaporated from the sea rises high in the air, it cools and condenses on dust particles to form tiny water droplets. Huge numbers of these form clouds and when they join together they form raindrops.

The particle theory explains condensation in this way. The particles in a gas possess a large amount of energy, which they use to move. If the particles are cooled, they lose some of their energy and slow down. If the gas is cooled sufficiently, the particles lose so much energy that they can no longer bounce off each other when they meet. The particles now slide over each other and form a liquid.

Figure 8.10 Breathing onto a cold window causes water vapour in your breath to condense.

Sublimation

A few substances can change from a solid to a gas, or from a gas to a solid, without forming a liquid. This process is called **sublimation**. Solid carbon dioxide, known as dry ice, sublimes when it is heated to −78 °C. It can be used on stage to produce a mist in the air when it warms up, at rock concerts, for example.

Sulfur is released as a gas by volcanoes and as it cools it sublimes to form solid sulfur around the volcano's opening or vent. This solid sulfur is sometimes called flowers of sulfur.

Figure 8.11 The sulfur gas has sublimed to form a yellow solid on the rocks near the opening of the volcano.

13 How is melting different from evaporation?

14 How is boiling different from sublimation?

15 How are condensation and freezing similar?

16 Make a diagram to show the states of matter and the processes that change them. Start by copying out Figure 8.12. Then add the words 'evaporating', 'melting', 'boiling', 'condensing', 'subliming' and 'freezing' to the appropriate arrows.

The changes between the states of matter

The changes between states of matter can be shown as a triangle with double arrows running between the corners, as in Figure 8.12.

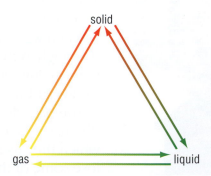

Figure 8.12 The interaction of the states of matter

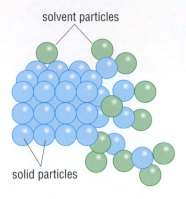

solvent particles

solid particles

Figure 8.13 Particles showing the dissolving process

Dissolving

When something **dissolves** in a liquid it forms a **solution**. The liquid is called a **solvent** and the solid that dissolves in it is called the **solute**.

The particle theory explains how things dissolve in the following way. There are small gaps between the particles in a liquid. When a substance dissolves in a liquid, its particles spread out and fill the gaps. Figure 8.13 shows how particles in a solid solute are pulled apart by the particles in the liquid solvent, which then move between them.

Figure 8.14 Copper sulfate dissolves in water to form a clear blue solution.

◆ SUMMARY ◆

◆ The three states of matter are solid, liquid and gas (*see page 122*).
◆ Each state of matter has its own set of properties (*see page 123*).
◆ The particle theory of matter explains what matter is made from (*see page 126*).
◆ The particle theory of matter can be used to explain the structure of the three states of matter (*see page 126*).
◆ Matter can change state in reversible reactions (*see page 126*).
◆ The particle theory can be used to explain how the structure of matter changes when it changes state (*see page 127*).
◆ The particle theory of matter can be used to explain how substances dissolve (*see page 131*).

End of chapter questions

1 Explain how the water in an iceberg may fall as rain over a city on the land.

2 Use the particle theory of matter to explain what happens to the particles when an ice cube melts and the water it produces evaporates.

- ◆ Matter and elements
- ◆ Metals and non-metals
- ◆ Different kinds of materials
- ◆ Properties of materials
- ◆ Property profiles
- ◆ Grouping materials according to their properties

In Chapter 8, we found out about the particle theory of matter. We saw that this theory can be used to explain the behaviour of matter as it changes from one state to another. We also discovered that the particles are atoms, or groups of atoms called molecules.

Introducing elements

Scientists use atoms to explain other things about matter. For example, a substance that has just one kind of atom is called an **element**. The atoms of each element are different from the atoms of all the other elements. Many substances are made from molecules, which are groups of atoms of one, two, three or even more elements.

Figure 9.1 This is a particle accelerator being prepared for investigating what happens when particles collide.

There are 93 elements that occur naturally. Another 24 elements have been made by scientists in laboratories, using specially designed apparatus called particle accelerators (Figure 9.1). Particles are whizzed round inside the accelerator, and then made to crash into each other to make atoms of new elements.

Metals and non-metals

As the number of elements is so large, scientists divide them into two groups – metals and non-metals. We are familiar with elements that are metals and can often recognise them by just one of their properties – they have a shiny surface.

Elements that are non-metals do not shine. They have dull surfaces. Two examples of non-metals are:

- carbon, which you might see as charcoal in a barbecue
- sulfur, a yellow substance that is used to make a wide range of substances from car tyres to medicines.

However, an element must have other properties if it is to be put in the metal group or the non-metal group. The properties that scientists use to classify elements into metals and non-metals are shown in Table 9.1.

Figure 9.2 Cooking pans are made of metal.

Figure 9.3 The surface of sulfur does not shine like the surfaces of metals.

Table 9.1 The properties of metals and non-metals

Property	Metal	Non-metal
surface	shiny	dull
physical state at room temperature	usually solid	solid, liquid or gas
strength	can be shaped by pressing and stretching without breaking easily	solids are usually soft or **brittle**
melting point	usually high	usually low
boiling point	usually high	usually low
density	usually high	usually low
conduction of heat	good	very poor
conduction of electricity	good	very poor

1 Name all the metals you know.

Most people have a good idea of what a metal is and can usually name a few different kinds. However, many people would find it difficult to identify a non-metal element even though we are surrounded by them. The air is a mixture of non-metal elements. Most of it (78%) is made up of nitrogen and a fifth (20%) is made up of oxygen. Chlorine is a non-metal element that is used worldwide to purify water by killing harmful microorganisms in it and make it fit for drinking. It is often used to keep the water clean in swimming pools. Another non-metal element that is widely used is phosphorus. Red phosphorus is used on the tip of matches and helps to produce a flame when the match is struck. As the phosphorus is rubbed against the matchbox, its temperature rises and the heat causes the non-metal to burst into flame.

Figure 9.4 Chlorine is used in many swimming pools to keep the water clean.

2 You are given a dull solid in the laboratory. How could you make some simple tests, without using heat, to see if it is a metal or a non-metal?

Figure 9.5 The smooth sides of some of these iodine flakes shine like pieces of metal.

Nearly all metals and non-metals have the properties shown in Table 9.1. For example, one property that *all* metals have is that they conduct electricity. You can test whether a material conducts using the electrical circuit shown in Figure 9.17 (page 143). If the sample conducts an electric current, the bulb lights up.

However, there are exceptions to the 'rules' described in Table 9.1. For example, some metals do not have a shiny surface – they are dull. Magnesium is an example of a metal that has a dull surface. When it is heated it catches fire and burns with a brilliant white flame.

Two exceptions from the non-metals are graphite and iodine.

- Graphite is a form of carbon, and is used to make pencils. It is a non-metal, but it conducts electricity.
- Iodine is used, with other substances, in some portable water-purifying kits. These kits are used by people who take adventure holidays and camp in places where the water may not be pure. Iodine has a shiny surface and looks metallic, but it is a non-metal.

Metal alloys

Not all the metals we use are single elements. Some of them are a mixture of metals called an **alloy**. Bronze is a mixture of copper and tin. It makes a ringing sound when struck and is used to make bells and cymbals.

Brass is an alloy of copper and zinc. It is strong and corrosion resistant, and is used to make the pins in electrical plugs. Its shiny surface makes brass a suitable metal for making ornaments.

Steels are very widely used alloys based on the metal iron. Carbon steels are made by mixing the metal iron with a small amount of the non-metal carbon. They are used to make many items, from springs to car bodies. Stainless steel is an alloy of steel and chromium. It does not rust as easily as other steels and is used to make knives, forks, spoons and kitchen sinks.

Figure 9.6 These bronze bells are used by Buddhist monks.

Atoms, molecules and everyday materials

We have seen that matter is made from the atoms of elements, and that elements can be divided into metals and non-metals. But what about everyday materials such as wood, cloth, pottery, glass and plastic? What are they made from? They are made from atoms of elements that are linked together to form molecules. Wood is made from carbon, oxygen and hydrogen atoms that have joined together to form a long-chain molecule called **cellulose**. Even a tiny piece of wood such as a fibre is made up from millions of cellulose molecules.

Figure 9.7 This is a model of part of a cellulose molecule. The grey balls represent carbon atoms, white balls are hydrogen atoms and red balls are oxygen atoms.

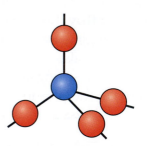

Figure 9.8 This is a model of the tetrahedrons made by silicon and oxygen atoms. The blue balls represents the silicon atom and the red balls represent oxygen atoms.

Cloth and plastic are also made from long-chain molecules, like wood. Clay and glass are made from silicon and oxygen atoms that join together to form tetrahedrons. These link together to form sheets in clay and a network in glass.

The arrangement of the atoms in materials affects their properties, but for now we do not need to know any more about this. However, we do need to know more about the properties of everyday materials.

Investigating everyday materials

People usually think of the word 'material' when they think about cloth, such as the clothes you are wearing now. But in science the word 'material' is used to mean any substance around you. The wood in a table or desk is a material. So is the water coming from a tap or the air around you. Humans have been selecting materials from the world around them from the earliest times and using the properties of these materials to help them survive. For example, the waxy coating of a large leaf has been used to make a waterproof shelter in a shower of rain for thousands of years.

Figure 9.9 These children in a rainforest are using leaves for shelter.

3 Imagine you have been cast away on a desert island (see Figure 9.10). What materials would you select to help you survive? Explain your answer.

Figure 9.10 What materials could you use to help you survive on a desert island?

Early materials – when and where?

When were some of the materials we use today first used? And where were they first used? Some materials – like stone, wood, the cane in baskets and the plant fibres in string and rope – have probably been used since the earliest times in the Stone Age. At the beginning of human history, it is likely that these materials were used everywhere in the world, although little evidence remains from those earliest times apart from some stone tools. Evidence for the use of other materials from later times is still being discovered by archaeologists today. The earliest evidence of the use of pottery has been found in China, dating from 18 000 years ago, while cave paintings at Lascaux, in France, show that humans were using paints about 17 000 years ago.

Figure A Stone was used for cutting and chopping in the earliest times.

Copper and gold are two of only a few metals that are found naturally in their metallic form – all the others are combined with other substances to form rocks known as metal **ores**. This meant that copper and gold were two of the earliest metals to be used. A piece of copper jewellery dating from about 10 000 years ago has been found in Northern Iraq. Evidence of gold being used to make ornaments has been found in Bulgaria dating back 6700 years but it may have been used for decoration well before that date.

About 8000 years ago, it was discovered that heating metal ores in a fire caused the metals to appear in the ash. This process is known as smelting. People found that when tin and copper were smelted from their ores and mixed together, they formed an alloy called bronze. This discovery led to the Bronze Age, which began in Canaan in the land between the eastern Mediterranean Sea and the river Jordan.

Meteorites made of iron have hit the Earth in the past and when people found them they may have used the metal to make tools. When it was discovered how to smelt iron, about 3200 years ago, the Iron Age began. Some of the earliest places where iron was smelted and used were in the countries now known as Turkey, India and China.

The first people wore animal skins but after the domestication of sheep, about 12 000 years ago, woollen clothing would have been gradually developed. The earliest evidence of woollen fabric, however, is only about 3500 years old and was found in Denmark. About 7000 years ago, people in countries from the Mediterranean to India grew the flax plant and used its fibres to make linen. The most extensive users of this material were the Egyptians.

At around the same time, in the area that is now Northwest India and Eastern Pakistan, cotton was grown and made into clothing and sheets. About 6000 years ago in South America, people were beginning to use cotton for clothing and in time this spread to North America and the West Indies. At around the same time, silk production first began in China and in time spread to Japan, Korea and India.

One of the last of the early materials we still use today is glass. This was invented in Mesopotamia (the region that is now Iraq) 5500 years ago. Other materials were discovered or invented much later. For example, paper was first used about 1900 years ago, while the first plastics – our most widely used materials today – were developed less than 200 years ago. In fact, most of the plastics we use today were developed in the last 80 years, and more are being developed in your life time!

Figure B This bronze statue was made in Benin in West Africa.

Figure C In Ancient Egypt, most clothing was made from linen.

For discussion

1 **Make a time line from 20 000 years ago to 5000 years ago. Mark on the time line when each of the materials discussed here was first used or developed.**

2 **Look around you and see if you can date any of the materials you see to their time in early history.**

3 **Use an atlas to find the places where the materials were first used or developed and mark them on a blank map of the world.**

1 Making careful observations, suggesting ideas that may be tested, choosing appropriate apparatus and using it correctly, and making conclusions from collected data, are all parts of scientific enquiry. How do you think the early metal workers could have used these activities in discovering bronze?

The properties of materials

There are a number of properties that materials may have. Here is a list of these properties, with some examples of materials that possess them.

- **Surface appearance.** Shiny, dull, rough or smooth.
- **Rigid.** A rigid material cannot be bent or squashed. Rock has been used as building material from early times because it is rigid.
- **Flexible.** A flexible material can be bent or squashed but when the pushing or pulling force is removed it springs back to its original shape. Flexible pieces of wood have been used for thousands of years for hunting bows and are still used today in some parts of the world (Figure B, page 11).
- **Hard or soft.** Most materials have a hard surface, which can be due to them being rigid. Materials such as sponge feel soft because of their flexibility.
- **Malleable.** A malleable material can be shaped by hammering or by pressing, without the material cracking. It stays in the shape after the shaping process has ended. Metals such as gold, silver and copper are malleable. They can be made into wires, and bent to form jewellery such as necklaces, bangles and earrings.
- **Brittle.** A brittle material breaks suddenly if it is bent or hit. You can snap a biscuit or a chocolate bar because it is brittle.

4 Look at the surfaces of the materials around you. Which are shiny, dull, rough or smooth?

5 What would happen to the castle in Figure 9.13 if the rocks suddenly lost their rigidity?

6 Is the shell of a hen's egg as brittle as the shell of a duck or goose egg? What investigation could you make to find out?

Figure 9.11 This castle in Aragón, Spain, was built of rock hundreds of years ago.

Figure 9.12 Gold is the most malleable of the metals used to make jewellery.

7 Devise an investigation to compare the absorbent properties of different brands of paper towels.

● **Absorbent.** An absorbent material has holes in its surface through which water can pass and also has spaces inside where the water can collect. Absorbent cloths and papers are used to wipe up spills in kitchens and laboratories. Some kinds of rocks, such as sandstone or limestone, are absorbent – there are gaps, called pores, between the rocky grains and water can pass through them or fill them up to make an underground store of water called an aquifer.

Oasis

An oasis is a spring in the desert around which plants grow. It forms because under the desert surface there is a layer of porous rock over the top of a layer of non-porous rock. The two layers stretch back from the desert under surrounding hills and mountains. When it rains in these places, the water drains into the porous rock but is prevented from sinking further by the non-porous rock below. The water then moves out along the layer of porous rock under the desert, forming an aquifer. In a place where the overlying surface is thin, the water may burst through forming a spring and a pool. The damp soil around the pool provides a habitat for plants, which in turn provide a habitat for animals.

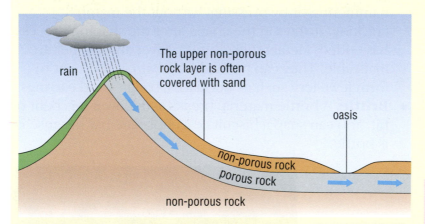

Figure D How water reaches the desert from the mountains and forms an oasis

Figure E An oasis in the Desert of Ica, Peru

2 Why can an oasis be a lifesaver to travellers lost in the desert?

3 Which property does the non-porous rock have that lets it stop water passing through it? Look at the list of properties on these pages to find out.

4 You are given two samples of rock from an oasis. Plan an investigation to find out which one prevents water sinking into it and which one allows water to pass through it.

Figure 9.13 Water-resistant material in an umbrella provides protection from the rain.

- **Waterproof.** A waterproof material does not let water pass through it. There are two kinds – water-resistant materials and water-repellent materials. A water-resistant material is made of fibres. There are holes between the fibres through which water could pass. However, the fibres are coated in **silicones**, which make the water gather up into large droplets that cannot pass through the material. Water-resistant materials are used to make umbrellas, outdoor jackets and trousers. A water-repellent material does not have any holes in it through which water can pass. Water-repellent material is used to make wellington boots, for example.

- **Transparent.** A transparent material is one that lets light pass through it without the light rays being scattered. This means that you can see objects clearly through transparent materials. You are looking through a transparent material at these words. Air is a transparent material, and so is water, although we tend to think of transparent materials as being solids such as glass and plastic.

- **Translucent.** A translucent material also lets light pass through it but the light rays are scattered. This means that you cannot see objects clearly. Translucent glass is used in bathroom windows.

- **Opaque.** Opaque materials do not let light pass through them. Most materials are opaque but one material that is used especially for its opaque property is curtain fabric. Curtains prevent light entering a bedroom on a sunny morning and prevent people outside seeing into homes in the evening. Opaque materials are also useful in making sunshades.

Figure 9.14 In countries with long periods of sunny weather, parasols and canopies provide shade.

Figure 9.15 As heat is conducted along the metal rod it glows red. The blacksmith holds the hot metal using a thick fabric glove – the fabric is an insulating material.

8 How could you find out which material is the best conductor of heat, using a metal spoon, a plastic spoon, a wooden spoon and a piece of aluminium foil, a bowl of hot water and some butter and a knife?

- **Heat conductor.** A material that is a heat conductor allows heat to pass through it. The particles from which the material is made pass heat energy along, from one particle to the next.
- **Heat insulator:** A material that is a heat insulator does not let heat pass through it, because its particles do not pass heat easily from one to the next. Good insulators are also known as bad conductors.
- **Electrical conductor.** A material that is an electrical conductor allows an electric current of electricity to flow through it.
- **Electrical insulator.** A material that is an electrical insulator does not allow an electric current to flow through it.

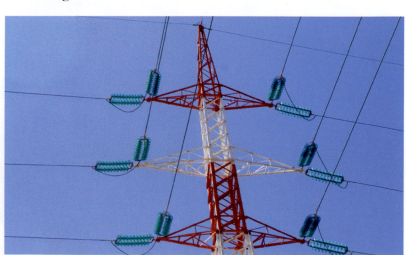

Figure 9.16 On this pylon, the metal cable conducts electricity between the power station and the city, while the glass insulators stop the current travelling to the pylon's metal frame.

Making a property profile

A description of a material in terms of its properties is called its property profile. You can make a property profile of a material by testing it for each of the properties in the list above.

When testing for flexibility it is important to consider the shape of the material. For example, a block of wood or plastic is not flexible, but a strip of the same material, such as a wooden or plastic ruler, is flexible.

A solid can be tested to find out if it conducts electricity by using a circuit like the one shown in Figure 9.17. The solid to be tested is secured between the pair of crocodile clips and the switch is closed. The bulb lights up if the solid conducts electricity.

By using this circuit, metals and the non-metal carbon, in the form of graphite, are found to conduct electricity. Other non-metals such as plastic and pottery do not conduct electricity.

> **9** Can metals, plastics and pottery be both rigid and flexible? Explain your answer.

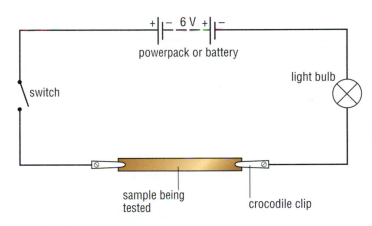

Figure 9.17 A circuit for testing conduction of solid materials

Comparing the properties of materials

Once you have made a property profile of a few materials, you can make a table like Table 9.2, on the next page, and see which materials could be grouped in the 'absorbent group', the 'brittle group' and so on.

Table 9.2 Comparing the properties of materials

Property	Material					
	A	B	C	D	E	F
appearance						
rigid						
flexible						
hard/soft						
malleable						
brittle						
absorbent						
waterproof						
transparent						
translucent						
opaque						
heat conductor						
heat insulator						
electrical conductor						

◆ SUMMARY ◆

◆ An element is made up of one kind of atom (*see page 132*).
◆ Elements can be divided into metals and non-metals (*see page 133*).
◆ Metals and non-metals have different properties (*see page 134*).
◆ There are a few metals and non-metals that have exceptional properties (*see page 135*).
◆ Some metals are mixtures of metals, called alloys (*see page 135*).
◆ There is a wide range of properties that a material might have (*see page 139*).
◆ Materials can be examined to find which properties they possess (*see page 143*).
◆ Materials can be assigned to groups according to their properties (*see page 143*).

End of chapter questions

1 Here are some items you may find in a kitchen:
   • metal saucepan   • wooden spoon   • glass measuring jug   • cotton cloth
   Here are some properties of materials:
   • absorbent   • good conductor of heat   • good heat insulator   • transparent
   a) Match each item with a property of the material from which it is made.
   b) How does this property of the material make the item useful when you are making a meal?

2 How many different materials are your shoes made from? What are the properties of each material? How are these properties useful?

10 Acids and alkalis

◆ The meanings of the terms 'acid' and 'alkali'
◆ Acids produced by living things
◆ Organic and mineral acids
◆ Uses of alkalis in the home
◆ Detecting acids and alkalis using indicators
◆ The pH scale
◆ Neutralisation
◆ Using neutralisation reactions
◆ Acid rain and its prevention

In Chapter 9, we looked at different kinds of substances and examined their properties. In this chapter, we are going to look at two kinds of solutions that are widely used in science. They are called acids and alkalis.

Early acids and alkalis

People have known since the time of the Ancient Egyptians and Greeks that some substances taste sour and some feel slippery. Vinegar is probably the best example of a sour-tasting liquid that early peoples would have known about. Early examples of slippery substances included potash found in the ashes of burnt wood, soda made from the evaporation of some **solutions**, and lime made from the burning of seashells. Scientists developed the word acid from *acidus*, which is the Latin word for 'sour'. The word alkali was developed from *al-qaliy*, which is an Arabic word meaning 'the ashes'.

A great deal of early investigative work in chemistry was done in Islamic countries, starting about 1200 years ago. Probably the greatest of the Muslim chemists at this time was Jabir ibn Haiyan who was also known as Geber. He worked on many investigations, which resulted in him devising new apparatus and discovering different kinds of acids.

1 Which pieces of apparatus shown Figure 10.1 are similar to those we use today? Look at Figure 7 on page 5 to help you answer.

2 How is the source of heat shown in Figure 10.1 different from the laboratory heat sources we use today?

Figure 10.1 Scientists in Geber's time showed their discoveries to others and built up a vast amount of knowledge about chemistry.

Acids

Most people think of **acids** as corrosive liquids that fizz when they come into contact with solids and burn when they touch the skin. This description is true for many acids, and when they are being transported the container holding them has the hazard symbol shown in Figure 10.2.

corrosive   **Figure 10.2** The hazard symbol for a corrosive substance

Some acids are not corrosive, and are found in our food. They give some foods their sour taste. Many acids are found in living things. Table 10.1 shows some acids found in plants and animals.

Table 10.1 Acids found in plants and animals

Acids with plant origins	Acids with animal origins
• citric acid in orange and lemon juice • tartaric acid in grapes • ascorbic acid (vitamin C) in citrus fruits and blackcurrants • methanoic acid in nettle stings	• hydrochloric acid in mammalian stomach • lactic acid in muscles during vigorous exercise • uric acid in urine • methanoic acid in ant sting

3 Look at Table 10.1 and state the organ systems in your body where three of the acids are found.

The acid in both ant stings and nettle stings is methanoic acid.

Grapes contain tartanic acid

Figure 10.3 Animals and plants that produce acids

The acid in vinegar

Ethanoic acid is found in vinegar and is produced as wine becomes sour. The wine contains ethanol and also has some oxygen dissolved in it from the air. Over a period of time, the oxygen reacts with the ethanol and converts it to ethanoic acid. This chemical reaction happens more quickly if the wine bottle is left uncorked.

Organic acids and mineral acids

The acids produced by plants and animals (with the exception of hydrochloric acid) are known as organic acids. Ethanoic acid is also an organic acid.

Mineral acids are not produced by living things and their discovery began with the work of chemists such as Geber. The first mineral acid to be discovered was nitric acid. It was used to separate silver and gold. When the acid was applied to a mixture of the two metals it dissolved the silver but not the gold. Later, sulfuric acid and then hydrochloric acid were discovered.

4 Why does wine go sour faster if the cork is removed from the bottle?

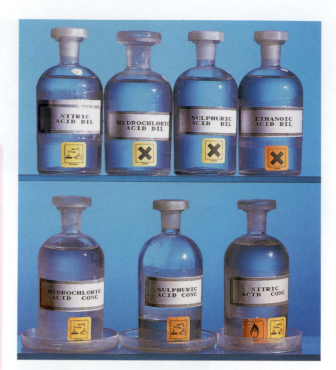

Figure 10.4 Bottles of dilute and concentrated acids

Alkalis

Sodium hydroxide solution and potassium hydroxide solution are examples of **alkalis** that are used in laboratories. Calcium hydroxide, also called slaked lime, is used in many industries to make products such as bleach and whitewash. A weak solution of calcium hydroxide is used in the laboratory, where it is known as limewater. It is used to test for carbon dioxide gas. If this gas passes through limewater, it turns the limewater milky.

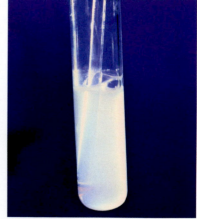

Figure 10.5 Limewater is clear but becomes cloudy when carbon dioxide is bubbled into it.

Figure 10.6 Alkalis used in the home

A concentrated solution of an alkali is corrosive and can burn the skin. The same hazard symbol as the one used for acids (Figure 10.2) is used on containers of alkalis when they are transported. Even dilute solutions of alkali, such as dilute sodium hydroxide solution, react with fat on the surface of the skin and change it into substances found in soap. Many household cleaners used on metal, floors and ovens contain alkalis and must be handled with great care.

7 Why should alkalis be treated with care?

8 What question do you think Boyle asked himself when he learnt about the work of the French dyers?

9 Which scientific skill did Boyle use in his tests on plant juices?

10 When Boyle tested a liquid with red cabbage juice, it turned the indicator from purple to red.
a) What was the liquid?
b) What colour would it turn Boyle's violet juice?

11 When Boyle tested a liquid with litmus solution, it went blue.
a) What was the liquid?
b) What colour would it turn violet juice?
c) What colour would it turn red cabbage juice?

Detecting acids and alkalis

Robert Boyle was an Irish scientist who lived just over 300 years ago. He studied acids and alkalis and decided to try and find an easy way to identify them. He knew that in France workers who made silk clothes dyed them with the juices of plants, and he began testing plant juices to see if they would solve his problem.

When Boyle tested acids and alkalis with the juice from red cabbage, he found a way to identify them easily. When acid is added to red cabbage juice, it turns from purple to red. When alkali is added, the juice turns from purple to green. He also found that juices from violets turned purple with acid and greenish yellow with an alkali, but his discovery about the colour change in litmus, a juice from a **lichen**, went on to be used in chemistry laboratories around the world.

Litmus is used as a solution, or it is absorbed onto paper strips. Litmus solution is purple but it turns red when it comes into contact with an acid. Litmus paper for testing for acids is blue. The paper turns red when it is dipped in acid or a drop of acid is put on it. When an alkali comes into contact with purple litmus solution, the solution turns blue. Litmus paper used for testing for an alkali is red. When red litmus paper comes into contact with an alkali, it turns blue.

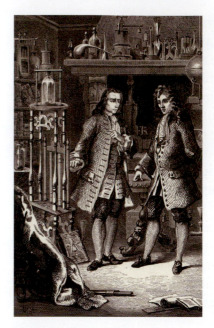

Figure 10.7 Robert Boyle and his assistant at work in his laboratory

> **12** Compare the laboratory of Geber (Figure 10.1) with that of Boyle (Figure 10.7).
> **a)** Do any pieces of apparatus in the different laboratories look similar?
> **b)** What differences do you notice?

There are over 20 indicators that scientists use. Here are two examples:

- methyl orange is pink in acid solutions and yellow in alkaline solutions
- phenolphthalein is colourless in acid solutions and pink in alkaline solutions.

There is even a plant that can be used as an indicator as it grows in the soil. It is the hydrangea. The colour of flowers can be affected by alkali in the soil. Hydrangeas have pink flowers when they are grown in a soil containing lime (calcium hydroxide, an alkali) and blue flowers when grown in a lime-free soil. The colour of the flowers can be used to assess the alkalinity of the soil.

Figure 10.8 Pink and blue hydrangeas

The pH scale

After indicators had been found to identify acids and alkalis, scientists wanted to know how to compare the strengths of acids and alkalis. In 1909, a Danish scientist called Søren Sørensen invented a scale called the **pH scale** to do just that. The letters p and H stand for the 'power of hydrogen' because this is an element that is found in acids, which takes an active part in their chemical reactions.

The pH scale runs from 0 to 14. On this scale, the strongest acid is 0 and the strongest alkali is 14. A strong acid has a pH of 0–2, a weak acid has a pH of 3–6, a weak alkali has a pH of 8–11 and a strong alkali has a pH of 12–14. A solution with a pH of 7 is **neutral**. It is neither an acid nor an alkali.

An electrical instrument called a pH meter is used to measure the pH of an acid or alkali accurately.

> **13** Look at Sørensen's laboratory in Figure 10.9. How is it different from the laboratories of Geber and Boyle?

14 Here are some measurements of solutions that were made using a pH meter:

solution A   pH 0
solution B   pH 11
solution C   pH 6
solution D   pH 3
solution E   pH 13
solution F   pH 8

a) Which of the solutions are:
   i) acids
   ii) alkalis?

b) If the solutions were tested with universal indicator paper, what colour would the indicator paper be with each one?

c) Fresh milk has a pH of 6. How do you think the pH would change as it became sour? Explain your answer.

15 Here are some results of solutions tested with universal indicator paper:

sulfuric acid – red
metal polish – dark blue
washing-up liquid – yellow
milk of magnesia – light blue
oven cleaner – purple
car battery acid – pink
Arrange the solutions in order of their pH, starting with the one with the lowest pH.

16 Identify the strong and weak acids and alkalis from the results shown in questions **14** and **15**.

17 Look at page 147 about acids and predict whether nitric acid is a strong or a weak acid. Explain your answer.

Figure 10.9 Søren Sørensen at work in his laboratory

For general laboratory use, the pH of an acid or an alkali is measured with universal indicator. This is made from a mixture of indicators. Each indicator changes colour over part of the range of the scale. By combining the indicators, a solution is made that gives different colours over the whole of the pH range (Figure 10.10).

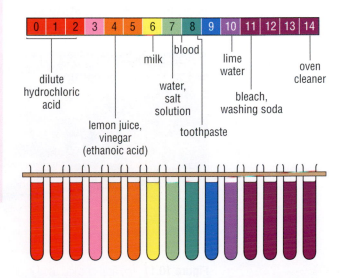

Figure 10.10 The pH scale (top) and universal indicator (bottom)

Neutralisation

When an acid reacts with an alkali, a process called **neutralisation** occurs in which a salt and water are formed. This reaction can be written as a general word equation:

$$\text{acid} + \text{alkali} \rightarrow \text{salt} + \text{water}$$

Specific examples of neutralisation reactions are:

$$\text{hydrochloric acid} + \text{sodium hydroxide} \rightarrow \text{sodium chloride} + \text{water}$$

$$\text{sulfuric acid} + \text{potassium hydroxide} \rightarrow \text{potassium sulfate} + \text{water}$$

$$\text{nitric acid} + \text{sodium hydroxide} \rightarrow \text{sodium nitrate} + \text{water}$$

Sodium hydrogen carbonate is a white solid. It is not an alkali but dissolves in water to produce an alkaline solution. It also takes part in neutralisation reactions with acids but produces another substance as well as a salt and water. It produces carbon dioxide. The word equation for this reaction involving hydrochloric acid is:

$$\text{sodium hydrogen carbonate} + \text{hydrochloric acid} \rightarrow \text{sodium chloride} + \text{carbon dioxide} + \text{water}$$

Sodium hydrogen carbonate is also called sodium bicarbonate. It has several uses in neutralisation reactions. Some of these are described in the next section.

Using neutralisation reactions

Insect stings

A bee sting is acidic and may be neutralised by soap, which is an alkali. A wasp sting is alkaline and may be neutralised with vinegar, which is a weak acid.

Figure 10.11 The acid or alkali used in an insect sting is delivered by a sharp point at the tip of the insect's abdomen.

Curing indigestion

Sodium bicarbonate is used in some of the tablets that are made to cure indigestion. Indigestion is caused by the stomach making too much acid as it digests food. When a tablet of sodium bicarbonate is swallowed, the chemical dissolves to make an alkaline solution, which neutralises the acid in the stomach and cures the indigestion.

Baking a cake

Baking powder contains a mixture of a solid acid and sodium bicarbonate. When the baking powder is mixed with water and flour to make a cake, the acid and the sodium bicarbonate dissolve in the water and take part in a neutralisation reaction. The carbon dioxide gas forms bubbles in the mixture and makes it rise to give the cake a light texture.

A model volcano

In the past, you may have made a model volcano. To do this, you may have added a tablespoon of sodium bicarbonate, called baking soda, to an empty plastic drink bottle and then built a mound of sand around the bottle so that it looked like a conical volcano. Finally you may have added red dye to half a cup of vinegar, then poured the vinegar into the bottle. Moments later a red froth would have emerged from the top of the bottle and flowed down the cone of sand, like lava flowing down a volcano (Figure 10.12). The model looks impressive! It does not illustrate how lava is formed, but it does show the power of a neutralisation reaction – between the baking soda and the vinegar.

Figure 10.12 The ingredients (left) for making a model volcano (right)

Fighting a fire

The soda–acid fire extinguisher contains a bottle of sulfuric acid and a solution of sodium bicarbonate. When the plunger is struck or the extinguisher is turned upside down, the acid mixes with the sodium bicarbonate solution and a neutralisation reaction takes place. The pressure of the carbon dioxide produced in the reaction pushes the water out of the extinguisher and onto the fire.

Improving crop growth

Acidity in the soil affects the growth of crops. It makes them produce less food. Lime (calcium hydroxide) is used to neutralise acidity in soil. When it is applied to fields it makes them appear temporarily white, as Figure 10.13 shows.

Figure 10.13 Liming fields in England to improve crop production

Acid rain

Natural acid rain

Water vapour high in the air condenses on dust particles to form huge numbers of tiny water droplets. They reflect light in all directions and we see them as a cloud. Carbon dioxide present in the air dissolves in the water and forms a weak acid called carbonic acid.

When this weak acid rain falls onto the rocks in limestone country, a chemical reaction takes place. In this reaction a very small amount of limestone is dissolved and washed away. Over thousands of years the dissolving of the rocks can produce a range of features in the landscape.

When part of a limestone surface is dissolved, cracks called grikes form and the surface is called a limestone pavement. In some places where most of the limestone surface is dissolved away, pinnacles of rock are left behind. As the acid rainwater dissolves the rock, it sinks into it and makes passages for underground streams. In time, the water passing along these streams dissolves more of the rock and creates caves. Eventually, as the cave becomes larger and larger, its roof may weaken and collapse to make a gorge.

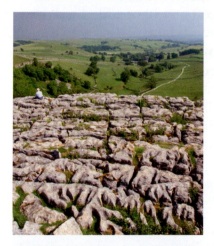

Figure 10.14 This limestone pavement is in the Yorkshire Dales, England.

22 Describe how naturally acidic rainwater can affect a mountain of limestone.
23 A sample of acid rain turned universal indicator yellow. What would you expect its pH to be? Is it a strong or a weak acid?

Air pollution and acid rain

Fuels such as coal and oil contain sulfur. When the fuel is burnt, the sulfur takes part in a chemical reaction with oxygen in the air and sulfur dioxide is produced. This gas reacts with water vapour and oxygen in the air to form sulfuric acid. This may fall to the ground in raindrops and make acid rain that is much more harmful than usual.

Power stations that burn coal and oil produce sulfur dioxide in the gases they release from their chimneys. They also produce chemicals called oxides of nitrogen, which form nitric acid and this too makes rain acidic. Oxides of nitrogen are also produced in the exhaust gases of cars and trucks.

The effect of acid rain

When acid rain reaches the ground, it drains into the soil, dissolves some of the minerals there and carries them away. This process is called **leaching**. Some of the minerals are needed for the healthy growth of plants. Without the minerals the plants become stunted and may die.

The acid rain drains into rivers and lakes and lowers the pH of the water. Many forms of water life are sensitive to the pH of the water and cannot survive if it is too acidic. If the pH changes, they die and the animals that feed on them, such as fish, may also die. Acid rain leaches aluminium ions out of the soil. If they reach a high concentration in the water, the gills of fish are affected. It causes the fish to suffocate.

24 In the Arctic regions, snow lies on the ground all winter.
As spring approaches and the air warms up, some of the water in the snow evaporates. Later, all the snow melts.
a) How does the evaporation of the water in the snow affect the concentrations of acids in the snow?
b) The table shows how the pH of a river in the Arctic may change during the spring.
  i) Plot a graph of the data.
  ii) Why do you think the pH changed in weeks 5–7?
  iii) Why do you think the pH changed in weeks 8–10?
  iv) How do you expect the pH to change in the next few weeks after week 10? Explain your answer.

Week	pH
1	7.1
2	7.0
3	6.9
4	6.8
5	5.5
6	5.0
7	4.7
8	5.1
9	5.5
10	5.9

Neutralisation and acid rain

Rivers and lakes that have become acidic due to acid rain can be treated with lime (calcium hydroxide). This reacts with the acid in the water and neutralises it. This method of restoring aquatic habitats has been used in Norway, Sweden and Wales. It is expensive and must be applied for as long as the source of the acid rain affects the rivers and lakes.

A better way is to remove the cause of the acid rain where it is being produced – at the power station. In the power station chimney a mixture of lime and water is sprayed into the smoke. Sulfur dioxide dissolves in the water and takes part in a neutralisation reaction with the lime. Calcium sulfate is produced. This substance is also known as gypsum. It is made into plaster of Paris, which is used to support broken limbs in hospitals and is also used in the making of cement and concrete in the building industry.

◆ SUMMARY ◆

◆ Some acids are made by living things (*see page 146*).

◆ The acid in vinegar is ethanoic acid (*see page 147*).

◆ The mineral acids are nitric acid, sulfuric acid and hydrochloric acid (*see page 147*).

◆ Sodium hydroxide and potassium hydroxide are examples of alkalis (*see page 148*).

◆ An acid turns blue litmus paper red (*see page 149*).

◆ An alkali turns red litmus paper blue (*see page 149*).

◆ The pH scale is used to measure the degree of acidity or alkalinity of a solution (*see page 150*).

◆ When an acid reacts with an alkali a neutralisation reaction takes place (*see page 151*).

◆ Neutralisation reactions have a wide range of uses (*see page 152*).

◆ Acid rain can form naturally (*see page 154*).

◆ Air pollution can cause acid rain (*see page 155*).

◆ Alkalis can be used to prevent acid rain (*see page 156*).

End of chapter questions

1 Write an account entitled 'The acids and alkalis in our lives'.

2 How can you tell when an acid has neutralised an alkali?

11 Rocks and soil

If you pick up a handful of earth what do you find? Rock and soil. How did they form? How are they related to each other? You can find out in this chapter.

Rocks

We live on a rocky planet that spins on its axis once a day and moves in an orbit around a star – the Sun – at 29 kilometres per second. How did the star, the Earth and its movements come about? Scientists believe it began with a huge explosion.

1 How far does the Earth travel in an hour?

From the Big Bang to the Sun

Scientists believe that the moment the universe came into existence it was an extremely hot, tiny white spot with an enormous mass. It expanded rapidly in an explosion that has been named the **Big Bang**, about 14 billion years ago. As the universe expanded, it cooled down and the atoms of the first elements formed. They were hydrogen and helium.

The atoms were drawn together by the force of **gravity** to form huge gas 'clouds'. In these clouds of gas the force of gravity brought some of the atoms even closer together until they formed huge spheres of hot gas. Inside each sphere the force of gravity drew the atoms closer still until the temperature and pressure became so great that a process called **nuclear fusion** took place. In this process some of the hydrogen atoms were converted into helium atoms and large amounts of energy were released. The energy escaped from the surface of the gas spheres as light and heat, and the spheres became what we now call stars.

Figure 11.1 Stars can be thought of as chemical factories where elements are made.

When a star has used up all its supply of hydrogen to make helium it makes other elements by nuclear fusion. These elements include carbon, nitrogen, oxygen, sodium, magnesium, aluminium, silicon, phosphorus, sulfur, chlorine, potassium, calcium, chromium and iron. The star may then swell up and form a red giant star, then release gas and dust as a **nebula** around it, and shrink to form a white dwarf star.

For discussion

The main elements in your body are carbon, hydrogen, oxygen, nitrogen, sulfur and phosphorus. They are joined together to make compounds. In some science fiction stories, terms such as 'people of the stars' or 'star children' are used. Is there any link between stars and people? Explain your answer.

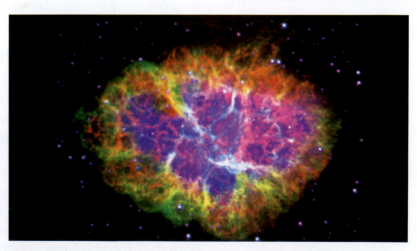

Figure 11.2 The remains of a supernova spreading out through space.

Very large stars do not release gas and dust gradually – they explode, forming a supernova, and shoot out gas and dust over a wide region of space. The conditions in these very large stars allow elements with larger atoms to be formed. These elements include nickel, copper, zinc, silver, tin, iodine, platinum, gold, mercury and lead.

The first stars that formed in the universe are called first-generation stars and the gases and dust (made from the newly formed elements) from these first red giants and supernovae formed a second set of stars called second-generation stars. These behaved just like the original stars and produced more of the same elements. The gas and dust from them also spread out through space. The next set of stars, the third-generation stars, is the set to which our Sun belongs.

The formation of the Solar System

Scientists believe that about 4.6 billion years ago the Sun and the Solar System formed from a huge cloud of gas and dust. They think an exploding star nearby made the cloud begin to rotate. As the cloud turned it formed a disc. Hydrogen and helium in the cloud collected at the centre and formed a star, our Sun.

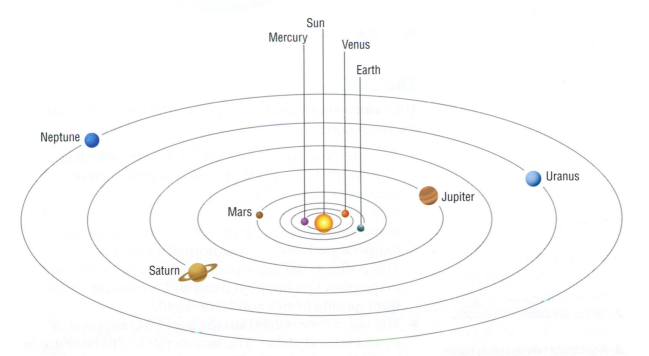

Figure 11.3 The Solar System (not to scale)

The dust particles moving in the disc around the Sun began colliding into each other and sticking together. Over time they formed larger and larger lumps of rock. Some became the first four planets moving around the Sun and others became the centres of four **gas giant** planets.

The structure of the Earth

The Earth beneath your feet is divided into three regions – the crust, the mantle and the core.

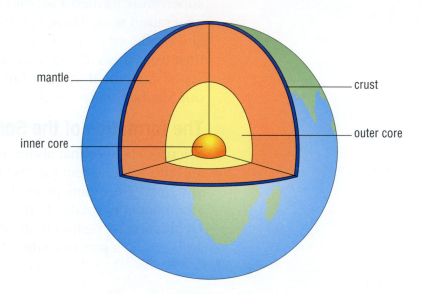

mantle — crust

inner core — outer core

Figure 11.4 The major parts of the Earth's structure

The core

The **core** is divided into two parts – the inner core and the outer core.

- The inner core is a ball of iron and nickel, which is 2740 km in diameter. There are also **radioactive elements** present, like uranium, and they generate heat, which keeps the core at about 5000 °C. The metals in the inner core still remain solid even though the temperature is above their normal melting point. This is due to the great pressure of the other materials in the planet pushing on them and preventing them from turning from a solid to a liquid.
- The outer core is 2000 km thick and is composed of more iron and nickel. The two metals in this layer are in liquid form. As the Earth turns, the inner core moves at a different speed from the outer core. This difference in movement in the two metallic regions is thought to generate the Earth's **magnetic field**.

2 Why is the centre of the Earth hot?

3 A compass needle points north when you let it move freely. Why do scientists think this happens?

The mantle

The **mantle** is made of rocky material and is 2900 km thick. It is composed mainly of the elements iron, silicon, oxygen and magnesium. The atoms of these elements are joined together to make substances called **compounds**. The main compounds in the mantle are called silicates. They are made from silicon and oxygen atoms, which combine with atoms of other elements.

The mantle is very hot – for example, it is 1500 °C at a depth of 2000 km below the Earth's surface. This is above the normal melting point of the rock, but the pressure of the materials above it keep the rock solid. The upper mantle near the crust is cooler and is under less pressure. This allows the rocky material to behave like a very thick liquid and it flows, a little like toothpaste does when you squeeze the tube gently.

The crust

The Earth's **crust** is made from much cooler rocks than the mantle. Although the rocks at the surface can feel cool or cold, miners and cavers can feel an increase in temperature as they go down into the Earth's crust.

4 How is the mantle different from the core?

5 How is the crust different from the mantle?

The rock cycle

Where do the rocks in the crust come from? How do the different types form? These questions puzzled scientists for a long time but as they looked at the different types of rocks and where they were found, they began to build up an answer. It is best shown as a diagram called the **rock cycle**. A simplified diagram is shown in Figure 11.6.

Figure 11.5 This cave is in Sichuan Province, China. As the caver goes deeper the cave will become warmer and warmer.

Figure 11.6 The rock cycle

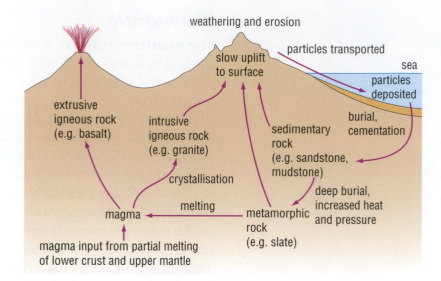

Scientists use diagrams to show how actions are related. The food chain on page 68 is a simple example. The rock cycle is more complicated but it can be understood by following the words and the arrows.

1 Find the word 'magma'. The arrow below it shows that this is molten rock that is made from some of the rocks in the lower crust and upper mantle.

2 Move up the arrow on the left. It shows that the magma becomes extrusive igneous rock. This means rock that escapes from the Earth's surface through a volcano, for example. Basalt (Figure 11.7) forms in this way.

3 From 'magma', move up the arrow on the right. It shows that magma forms intrusive igneous rock. This means that the rock cools down inside the Earth's crust. Granite (Figure 11.7) forms in this way.

4 Moving up the next arrow, you learn that a slow uplift of the rocks occurs to bring them to the surface. After this has happened you can see that **weathering** and **erosion** take place.

5 The arrow sloping down the mountainside to the right shows that the particles made during weathering and erosion go to the sea, where they are deposited on the seabed, as **sediments**.

6 The arrow coming from the seabed through the crust shows that the sediments become buried and the pieces of rock cement together. They form sedimentary rocks such as sandstone (Figure 11.11) and mudstone.

7 The short downward arrow shows that sedimentary rocks can be buried deep, where they are squashed and heated. These processes lead to the formation of metamorphic rock such as slate (Figure 11.14).

8 Upward pointing arrows from 'sedimentary rock' and from 'metamorphic rock' show that these rocks can be uplifted to the surface, like igneous rocks.

9 The arrow leaving 'metamorphic rock' and pointing to the left shows that these rocks can be melted to form magma.

Note that this simplified diagram shows only sedimentary rock forming metamorphic rock, but igneous rock can form metamorphic rock too. However, the arrows and labels would make the diagram more complicated and more difficult to interpret.

Types of rocks

As we have seen from the rock cycle, there are three types of rocks – igneous rocks, sedimentary rocks and metamorphic rocks. In this section we will look at these in more detail.

Igneous rocks

Igneous rocks are also called fire rocks and they form from magma. There are two kinds.

- Extrusive igneous rock, like basalt, reaches the Earth's surface.
- Intrusive igneous rock, like granite, forms within the crust.

Both rocks are made from crystals of **minerals** that stick together, but basalt is formed from small crystals while granite is formed from large crystals. This is because of the way the hot rocks cool down. Basalt, exposed to the air, cools down quickly, which makes small crystals form. Granite, remaining in the crust, stays warmer for longer and cools down more slowly. This makes large crystals form.

Figure 11.7 Basalt (left) and granite (right) – the crystals in the basalt are almost too small to be seen easily but the crystals of the different minerals in the granite can be clearly seen.

6 Describe how substances in granite could enter the magma.

Rapid cooling

Some rocks cool down so rapidly when they emerge into the air that crystals do not have time to form. Obsidian forms in this way. It is a black, glassy rock. Pumice forms from a frothy lava. The froth is full of bubbles made from gases formed in the hot rock. When the hot rock cools and the gases escape it turns into a very porous rock. It contains so many air spaces that it can float.

Figure 11.8 Obsidian (left) and pumice (right)

Volcanoes

The first rocks to be identified as igneous rocks were those that formed from volcanic eruptions. These have been taking place since the Earth first formed and continue today.

There are two kinds of volcanoes. The first type builds up pressure and then erupts in a huge explosion, sending rocks and dust high into the atmosphere. These volcanoes form cones and are known as andesitic volcanoes after the Andes mountains in South America where they were first identified. When the lava cools on the sides of a volcano like this, it forms a rock called andesite.

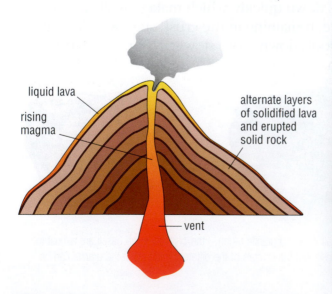

liquid lava

rising magma

alternate layers of solidified lava and erupted solid rock

vent

Figure 11.9 Structure and eruption of an andesitic volcano

The second type of volcano is called a basaltic volcano because of the type of rock it produces (Figure 11.7). It is flatter than the cone-shaped andesitic volcano and is sometimes called a shield volcano. Eruptions from basaltic volcanoes are usually much less violent than those of andesitic volcanoes and scientists can get quite close to study them. Steady and continuous basaltic eruptions occur under the sea, where the basalt forms the sea floor. They also occur on land on the Hawaiian islands.

Figure 11.10 A basaltic volcano erupting in Hawaii, USA

7 Why do you think igneous rocks are called fire rocks?

Sedimentary rocks

There are three kinds of **sedimentary rocks** – rocks that form from rocky fragments, rocks that form from parts of living things and rocks that form from minerals when seawater dries up.

Rocks from rocky fragments

Rocky fragments are produced by weathering and form small particles like sand grains, or even smaller. They are carried down rivers and settle out when the water current slows down. Over long periods of time – thousands or even millions of years – the layers of particles build up. As the thickness and the weight of the layers increases, the particles become squashed together. In time, the different particles become bound together by the minerals (page 169) that settle between them and make firm rock.

Sandstone is a good example of a sedimentary rock formed in this way. Other examples are rocks called conglomerates made with pebbles, shale made from a mixture of **silt** and clay, and siltstones and mudstones.

Figure 11.11 Sandstone is formed when tiny grains of rock are pressed and bound together.

Figure 11.12 You can see the shells in this piece of limestone.

Figure 11.13 Rock salt is also called halite.

8 Compare the ways in which the different kinds of sedimentary rocks are formed.

Rocks from parts of living things

The shells of ancient living things have formed sediments that have turned to rock. Limestone is formed from the shells of sea creatures such as molluscs (page 109) that collected at the bottom of the sea. Chalk forms from the tiny shells of protoctists (page 60) that lived in the **plankton**.

Rocks from dried-up seas

Seawater contains many chemicals. They have dissolved out of the minerals in the rocks and been washed down rivers into the seas and oceans. In the past many seas have dried up. In this process water evaporates from the surface, leaving the minerals behind to make the remaining seawater more concentrated. Eventually there is too little water left for all the chemicals to remain in solution, and some of them join together and form crystals. Rocks that form in this way are called evaporites. Rock salt and gypsum are two examples.

Metamorphic rocks

Metamorphic rocks are formed from igneous and sedimentary rocks that have been heated or squashed in the Earth's crust. The rise in temperature and pressure cause the rocks to change their form, or metamorphose. When limestone is heated and squashed, it changes into marble. Shale is a sedimentary rock made from very tiny particles similar to those in clay and mud. When it changes or metamorphoses into slate, the tiny particles or grains line up and make sheets of rock that can be split apart.

9 What does 'metamorphose' mean?

10 What causes a rock to metamorphose?

Figure 11.14 Marble (left) and slate (right)

Uses of rocks

Rocks have been used by people from the earliest times – the Stone Age was a huge length of time in human history. People selected rocks for different tasks, after examining their properties. Today, rocks still have many uses, as these examples show.

Uses of igneous rocks

Granite and basalt are very hard and so are used to make foundations for buildings and surfaces for roads. They are also used to make concrete for all kinds of buildings. When the surface of granite is polished, the different shapes and colours of its crystals give it an attractive surface. This makes it an attractive material to use for decorative stonework at the entrances to important buildings such as city libraries and museums.

Basalt is used to form protective shielding in nuclear power stations, which prevents the escape of harmful radiation.

Obsidian can be broken to produce a very sharp cutting edge and was used by Stone Age people to make knives and arrowheads. It has even been used today to make sharp knives called scalpels for surgeons.

Pumice was used to make concrete in Roman times and is still used to make lightweight concrete blocks today. It also breaks up easily when rubbed and can be used to clean the skin. Some toothpastes contain pumice to rub on the teeth and clean them.

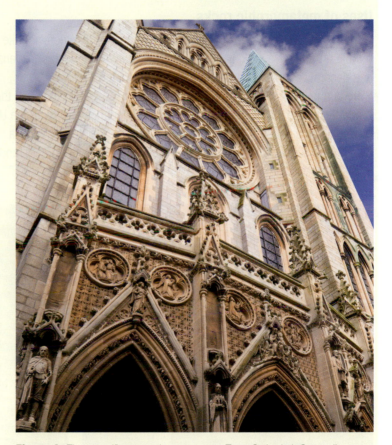

Figure A The magnificent granite entrance to Truro Cathedral, Cornwall, England

2 What property of granite makes it useful for:
a) making roads
b) forming the doorway to an important building?

Uses of sedimentary rocks

Sandstone and limestone form layers called beds. The place where the different beds meet is weaker than the surrounding rock. This allows the rock to be easily broken up into pieces that can be used as building stone. The pyramids of Egypt are made from limestone blocks. Limestone chippings are used to make roads, while more finely ground limestone is used to make cement and even some toothpastes. Limestone has many uses in the chemical industry. Rock salt is also used in the chemical industry, and in countries with very cold winters it is used for spreading on the roads. It dissolves in the water on the road and lowers the freezing point, which prevents the formation of ice on which traffic could skid. Gypsum is used in making cement and plaster.

Uses of metamorphic rocks

Its glistening, sugar-like texture and streaks of colourful minerals make marble an attractive rock. It is used to make statues, the tops of expensive and decorative tables and in the making of important buildings.

Slate is a non-porous rock and forms lightweight sheets. In the past, these properties made it a useful roofing material and in many towns in the UK it can still be seen on the roofs of houses. Sheets of slate also have a very smooth surface, and this rock is used to support the covers of pool and snooker tables.

3 Why are sandstone and limestone frequently used for building stone?

4 Limestone is a grey rock with a rough powdery texture. What does it become after it has metamorphosed? What properties does the new rock have?

5 How do the properties of slate make it useful for a roofing material?

Figure B The Taj Mahal in India is built from blocks of marble.

Figure C These houses in Lancashire, England are built from a sedimentary rock called millstone grit and have roof tiles made of slate.

Minerals

When researching about rocks you will often come across the term 'rocks and minerals' and perhaps you may wonder what is the difference between them. A **mineral** is a substance that has formed from one or more elements in the Earth. Rocks are formed from different minerals that stick together. This is shown well in granite (Figure 11.7). It is made from the minerals feldspar (pink), mica (black) and quartz (white).

Gold, silver and copper are examples of the very few elements that are found on their own. Most other minerals are compounds, made from the atoms of two or more elements that have joined together. Table 11.1 shows the ten most common elements in the Earth's crust.

Table 11.1

Element	% in crust
oxygen	50
silicon	25.8
aluminium	7.3
iron	4.2
calcium	3.2
sodium	2.4
potassium	2.3
magnesium	2.0
hydrogen	1.0
titanium	0.4

Figure 11.15 Tiny pieces of gold are separated from sediments in a process called panning. The water is swirled around in the pan to remove the sediment and leave the gold behind.

The atoms of the elements in a mineral usually join together to form a crystal structure. For example, the atoms of silicon and oxygen form a crystal substance called quartz, which can have different appearances, as Figure 11.16 shows.

Figure 11.16 Varieties of quartz – rock crystal (left) and amethyst (right)

Each mineral can be recognised by observing its crystal shape, colour, **lustre**, hardness and the colour of the streak it makes when it is rubbed across a rough, white porcelain surface. Over 2000 minerals have been identified.

Some rarer minerals have particularly attractive properties. They have a pleasing colour, a shiny surface or sparkle when light passes through them. These minerals – such as opal, diamond and beryl (which is cut to form emeralds) – are called gemstones. Different gemstones may be formed in different ways. For example, diamond (which is composed of just carbon atoms) forms in hot rock that rises through the Earth's crust. Beryl (which is composed of atoms of beryllium, aluminium, silicon and oxygen) forms in the last part of granite rock to cool in the Earth's crust. Opal (formed from silicon, oxygen and hydrogen atoms) forms from the minerals in the water of hot springs or from the weathering of certain kinds of rocks.

Ores

During the formation of some rocks, particularly sedimentary rocks, metal elements gather in large quantities. They are usually combined with the atoms of other elements. Rocks that possess large quantities of metal compounds are called **ores**. Bauxite is an ore rich in aluminium and haematite is an ore rich in iron. Ores are mined and then processed, usually using heat, to release the metal from its compounds in the rock.

11 What is a mineral?
12 What are the properties of gemstones?
13 Are gemstones minerals? Explain your answer.

Figure 11.17 Aluminium ore (bauxite) and iron ore (haematite)

Soil

We now know that the Earth is a rocky planet but if we stand on a natural piece of ground we find we are standing on soil. The soil is usually covered by plants that are growing in it. Soil allows plants to set down their roots and hold their position in a habitat. It also stores water and minerals for the plants to use as they grow. Successful farming depends on the soil, so over the years scientists have studied soil to see how it can best be used to grow crops.

How soil forms

The main part of a soil is made from small particles or fragments of rock. They are made when larger rocks break up. This process in which rock breaks up to form fragments is called **weathering**. There are two kinds of weathering – physical weathering and chemical weathering. Weathering can occur on rocks sticking out of the ground or on rocks under the soil.

Physical weathering of rocks above ground

Physical weathering is caused in four ways – by changes in temperature, by the effect of ice, by abrasion and by the effect of plants.

- **Changes in temperature**. When a rock heats up, the minerals in it expand. The minerals expand by different amounts and, as they do so, they push on each other. When the rock cools down, the minerals contract and spaces develop between them. After being heated and cooled many times the minerals become loose and fall away. In a desert, like the Sahara, the temperature soars during the day to over 40 °C and at night may drop to −7 °C. These great changes in temperature make the surfaces of rocks crumble.

Figure 11.18 This farmer in northern Peru is growing yams in a deep, crumbly, well-drained soil. These soil features are ideal for his crop.

14 a) The rocks around a campfire have a crumbly surface while other rocks in the area have smoother surfaces. Explain these observations.

b) When a campfire rock is split open, the inside has a smooth surface and the grains are tightly packed. Why?

Figure 11.19 This rock is shedding its outer layers as a result of many large changes in temperature.

15 Can ice cause weathering on any kind of rock? Explain your answer.

16 As a pebble is carried down a long river what happens to its size? Explain your answer.

- **The effect of ice**. When it rains on porous rock like sandstone (see page 165), the water can stay in the rock for some time. If the weather becomes colder, the water in the rock may freeze. When water freezes and forms ice, it expands. The ice pushes on the sides of the pores and makes the rock crumble. Water that gathers in the cracks of any rock also expands when it freezes. This makes it push on the sides of the crack and can cause pieces of rock to snap off.

- **Abrasion**. This occurs when pieces of rock rub together – when pebbles are carried in a fast-flowing river, for example, or thrown together at high tide on a beach in a storm. Glaciers, slowly moving down mountains, carry boulders that rub on the rocks in the valley floor and sides, and wear them down. Sand grains are fragments of rock. When they blow in the wind, in a sandstorm for example, they rub on rock surfaces and wear them down.

- **The effect of plants**. Particles of rock blown by the wind settle in cracks in rocks and form a soil in which plant roots can grow. If the seed of a tree or bush lands in a crack and germinates, the root can grow in the soil. Trees and bushes have strong, woody roots that grow larger every year, so if they grow in a crack they push on its sides and break up the rock.

Figure 11.20 In this cliff wall you can see the rock on which trees are growing. The tree roots are growing in cracks in the rock.

Chemical weathering of rocks above ground

Chemical weathering is caused in two ways:

- **Rainwater** is naturally slightly acid due to carbon dioxide from the air dissolving in it (see page 154). The acid in rainwater falling on rocks such as granite can break it up into smaller particles.
- **Hot and wet weather conditions** speed up chemical reactions such as those that break down rocks. This means that in places with hot, wet weather conditions, such as rainforest regions, rocks break down more quickly than in places with a cold, dry climate, such as the interior of the Antarctic continent.

17 What kind of weathering would you expect to take place in a desert?

18 Name a region of the world where you would expect weathering of soil to be fast. Explain your answer.

19 Name a region of the world where you would expect the weathering of soil to be slow. Explain your answer.

The fate of rocky fragments

When tiny rocky fragments have formed, they are usually carried away by water and sometimes by the wind. Eventually, when the water or wind slows down, the fragments settle out and form a layer, which can become a soil.

Figure 11.21 The rocky fragments carried down the river Nile have settled out and formed a delta where the river meets the Mediterranean Sea. A rich soil has formed which has been farmed for over 5000 years.

Chemical weathering of rocks below the soil

20 Could any kind of physical weathering take place at the bedrock? Explain your answer.

The layer of soil formed by rocky fragments lies on top of rock. This rock is known as **bedrock**. It also becomes weathered and its fragments join the other rocky fragments in the soil.

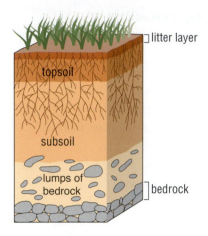

Figure 11.22 A side view of a soil, like this, is called a soil profile.

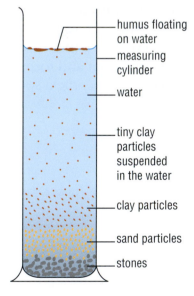

Figure 11.23 The separation of humus and the rocky components of a soil

When rainwater strikes the surface of the soil, it drains into it and moves down between the soil particles until it reaches the surface of the bedrock. As it moves down through the soil, some substances in the soil may dissolve in it. When the water and the dissolved substances arrive at the surface of the bedrock, weathering can take place and the pieces of rock that are produced can mix with the soil above it (Figure 11.22).

The soil layers

If you dig down into the ground, you can see a side view of the soil and discover that it is divided into layers.

At the surface is the darkest layer. It is made mainly from the remains of dead plants and animals. Bacteria in the soil feed on these remains and break them down. This breaking-down process produces a range of chemical substances and some of them dissolve in the rainwater, make it more acidic, and speed up the weathering of the bedrock. As the remains of plants and animals rot, they form a substance called **humus**.

The thicker layer under the surface is a mixture of humus and rocky fragments and is called the **topsoil**. It is the part of the soil that farmers and gardeners **cultivate** to make it suitable for growing their plants. It is also the layer that ecologists examine when they are studying a habitat.

Below the topsoil is the **subsoil**. This layer is paler than the topsoil because it contains much less humus.

Below the subsoil is a layer containing lumps of bedrock.

The parts of a soil

There are two parts to the soil – the humus and the rock fragments. There are four size groups for the rocky fragments. They are:

- small stones, over 2 mm across
- sand, 0.05–2 mm across
- silt, 0.002–0.05 mm across
- clay, under 0.002 mm across.

The amounts of the different rocky fragments in a soil can be compared by mixing up a sample of the soil with water in a tall clear container, letting it settle and then measuring the depth of each layer. In the stirring process, the humus separates from the rocky components and floats on the surface of the water.

The properties of a soil

There are four properties of a soil that are easy to investigate. They are texture, drainage (the ability to hold water), the amount of air and the soil pH.

Soil texture

The soil texture is a description of how the soil feels when you rub it between your finger and thumb. Sandy soil feels gritty, silty soils feel silky and clay soils feel sticky and cannot be squeezed as well as the other soils.

Drainage and water holding

The passage of water through the soil is called drainage. The drainage of soils can be compared in the following way. Dry samples of the same amount of each soil are set up in the apparatus shown in Figure 11.24.

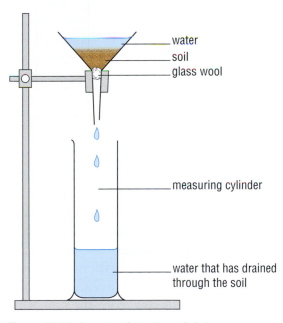

water
soil
glass wool

measuring cylinder

water that has drained through the soil

Figure 11.24 Apparatus for testing soil drainage

The same amount of water is poured onto each soil sample and the water is allowed to drain for the same amount of time. At the end of this time, the volume of water in each measuring cylinder is recorded to compare the drainage.

The investigation can also be used to compare the amount of water that each soil held back. This gives an indication of each soil's water-holding capacity. To find the volume of water held back, the volume of drained water is subtracted from the volume that was first added to the soil.

24 How does a soil feel that has particles between 0.002 and 0.05 mm across?

25 A soil is hard to squeeze. What size are its particles?

26 What would you predict the results to be if a sandy soil and a clay soil were compared for drainage and water-holding capacity? Explain your answer.

Air in the soil

The spaces between the soil particles contain air. This allows the plant roots and the soil organisms to respire (page 31). When water drains through a soil it flows through the air in the space.

If the air spaces are made to fill with water, the amount of air space in a soil can be found. The following investigation shows how this can be done.

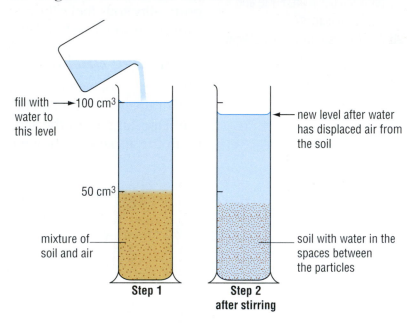

Figure 11.25 How to find the volume of air in a soil

Soil is added to a measuring cylinder until $50\,cm^3$ has been added. Water is poured in on top of the soil until the level of the water reaches the $100\,cm^3$ mark on the scale.

The soil is then stirred to help the water enter the air spaces. Stirring should continue until no more bubbles of air rise from the soil. When this happens, the water level should have fallen and its new position is noted on the scale. The volume of water now in the soil is found by subtracting the second water level from the first. This value is also the volume of the air that was in the soil at the beginning of the investigation. The percentage of air in the original volume of soil can be found by dividing the volume of air by the volume of the soil sample and multiplying by 100.

27 $50\,cm^3$ of soil was found to contain $10\,cm^3$ of air. What is the percentage of air in the soil?

28 Would you expect a clay soil to have more air in it than a sandy soil? Explain your answer.

Figure 11.26 Soil test kits are used by gardeners to find the pH of the soil.

29 Why is it best to use a *white* card, bowl or spoon when testing with the universal indicator?

Soil pH

The pH of a soil is a measure of how acidic or alkaline the soil is. It can be found by using universal indicator in either of the following two ways.

A sample of soil is shaken up with some universal indicator solution in a test tube and then allowed to settle. When this has happened, a white card can be held up behind the test tube to help you see the colour of the indicator. The colour indicates the pH of the soil.

A second method, which is especially useful with clay soils, is to mix the soil with water and a compound called barium sulfate, which is a white powder. The barium sulfate does not affect the acidity or alkalinity of the soil but binds clay particles together to make the water clearer for testing with the indicator. The water is drained from the soil and held in a white plastic spoon or bowl. Universal indicator is then added and its colour is observed to find the pH of the soil.

Types of soil

We have seen that weather affects the way a soil develops but there are other things that affect them too.
For example, the position of the soil in the landscape is important. If it is developing on a mountainside, it may have only thin layers and be composed of mostly rocky fragments, but if it is in the valley it may have much thicker topsoil and more humus in it.

In some regions in Africa, the rocks are rich in iron minerals and when they form soils the iron minerals make them red. In **tropical** regions that have wet and dry seasons, such as in India, weathering of the rocks produces soils rich in clay. They have cracks in their surface in dry weather but after rain the soil swells out to fill them.

30 Why do you think soils are:
   a) thin on steep mountainsides
   b) thicker in a valley?
31 After a dry season, when the rain begins, do you think water falling on clay will reach the bedrock quickly or slowly? Explain your answer. Why could the speed of reaching the bedrock change as the rains continue?

Figure 11.27 In this region of Leh in northern India, there are thin soils on the steep mountainsides and thicker soils in the valleys where crops are grown.

32 Which of these pH values might you find when testing soils in which rhododendrons, blueberries and ferns are growing – 5, 6, 7 or 8?

Soils have a low pH (are acidic) if they form on granite and have a high pH (are alkaline) if they form on limestone. There are other factors that affect the pH of the soil too. In places with a heavy rainfall, the water may wash alkaline substances out of the soil and make it more acidic. Sandy soils drain well and as the draining water takes alkaline substances from them, they become acidic. Clay soils drain poorly, so alkaline substances remain in them and make the soil alkaline.

Most plants are adapted for growing in a neutral soil or slightly acidic soil of pH 6. But some thrive in soils that are more acidic. Examples of plants that grow well in acidic soils are rhododendrons, blueberries, pines and ferns. Examples of plants that grow well in alkaline soils are oregano, asparagus, beech, lilac and sagebrush.

Loam

A soil with a large amount of sand is simply called a sandy soil. One with a large amount of silt is a silty soil, and a soil containing a large amount of clay is a clay soil. However, if a soil is made of 40% sand, 40% silt and 20% clay, it is called a **loam**. A loam also contains large amounts of humus and this binds the fragments together to make soil crumbs.

Figure 11.28 The crumbs in a loam

As the crumbs are larger than the rocky fragments, they have larger air spaces between them and this helps plant roots to respire and to grow down through the soil. The humus soaks up some of the water like a sponge and holds it there with dissolved minerals for the plant roots to take up. The mixture of sand and clay means that the soil has quite good drainage due to the sand, which is essential in wet weather, but also can hold some water due to the clay, which is useful when there is little rain. These features make the soil ideal for seedlings to grow in and for crop plants to produce a high yield of food. A loam is the soil that farmers and gardeners aim to cultivate. They do this by adding extra sand, clay, silt and humus to the soil until each is present in its correct percentage and the pH is at the correct value for the plants they wish to grow.

33 A soil has 70% sand, 20% silt and 10% clay.
   a) What kind of soil is it?
   b) How do its proportions need to be changed to make it into a loam?

◆ SUMMARY ◆

◆ The Earth formed from lumps of space rock that stuck together (*see page 159*).

◆ The structure of the Earth is divided into the core, the mantle and the crust (*see page 160*).

◆ The rock cycle shows how rocks form and move in the crust (*see page 161*).

◆ Igneous rocks form from molten rock inside the Earth (*see page 163*).

◆ Sedimentary rocks form from the settling down and compressing of materials (*see page 165*).

◆ Metamorphic rocks form from other rocks when they are squashed and heated in the Earth's crust (*see page 166*).

◆ Rocks are formed from minerals (*see page 169*).

◆ Rocks with large amounts of metal compounds in them are called ores (*see page 170*).

◆ Soil forms by the weathering of rock above ground (*see page 171*), and from the weathering of bedrock, which is below ground (*see page 173*).

◆ Soil is a mixture of rock fragments and humus (*see page 174*).

◆ There are four size groups of rocky fragments in a soil (*see page 174*).

◆ The soil has a number of properties such as texture, drainage and water-holding capacity, air content and pH (*see page 175*).

◆ There are different types of soil in different regions of the Earth (*see page 177*).

◆ Loam is a type of soil that many farmers and gardeners need for good plant growth (*see page 178*).

End of chapter questions

1 A building has a slate roof, granite walls, a marble floor and a sandstone doorway.
What processes took place to form these different kinds of building materials?

2 You are given a bucket of sandy soil, a bucket of clay soil and a bucket of loam.
a) What tests would you make on the soils to compare them?
b) What results might you expect for each kind of soil?

◆ How rock layers formed
◆ Naming the rock layers
◆ How fossils form in rocks
◆ Fossils and rocks
◆ The fossil record
◆ The fossil record and the ages of rocks
◆ Finding the ages of the rocks

The **landscape** is forever changing. Most of these changes take place far too slowly for us to notice. Occasionally some changes do take place quickly. For example, a volcano erupts and its lava and ash cover the ground, or a river floods and washes away its banks. These changes have been observed by people in the past and made them wonder how long they have been taking place and how old the Earth must be.

Figure 12.1 This landscape in Utah, USA, is gradually changing as the river erodes the rock. You can clearly see the layers in the sedimentary rock – the oldest are found at the bottom of the canyon.

Early ideas about rocks and the land

The Ancient Greeks discovered fossil shells of sea animals high in the mountains. This led one of their teachers, called Xenophanes, who lived about 2500 years ago, to suggest that at one time the mountains must have formed the sea floor and have been covered with seawater for the shells to settle on them.

Over 1400 years later, three scientists around the world were observing the landscape and trying to explain what they saw.

Al Beruni, a Persian with a wide interest in the world around him, looked at the ways **sediments** were laid down in the mouths of rivers entering the Indian Ocean. He concluded that at one time India was under the sea and was formed by the rocks settling out and replacing the sea with land.

Figure A The layers of sediments in a river bank

Avicenna, a Persian scientist, thought that processes he saw taking place around him in the landscape had also taken place in the same way in the past. He believed that if this was so, layers of rock seen in cliffs were formed like the sediments that were settling in rivers in his time. This meant that the layers at the bottom of a cliff were older than those at the top.

Shen Kuo, a scientist who lived in China, concluded from his observations on the landscape that rocks in the mountains and hills became worn down by weathering over a very long time and that this made the appearance of the landscape change. He also found fossils of bamboo in an area where in his time the bamboo could not grow. This made him believe that the **climate** in the area had also changed over a long period of time.

1 Can you think of another way in which the shells might have got up the mountains, without the mountains being part of the sea floor?

2 How do the observations and the ideas of Al Beruni, Avicenna and Shen Kuo suggest that the Earth may have changed?

How rock layers formed

You can see in the diagram of the rock cycle (Figure 11.6, page 162) that rocks are formed in three ways. They are formed by volcanic processes, by movements inside the Earth that heat and squeeze them and then raise them up, and by older rocks weathering and making sedimentary rocks. The layers of rock being examined in this chapter are sedimentary rocks. Each layer took millions of years to form.

Naming the layers of rock

When scientists realised that the rock layers could help them understand changes in the Earth's crust, they gave them names. Most of the deeper layers were named after an area where they form a major part of the landscape. The name also came to stand for a period of time in the Earth's history. The names of the two most recent layers of rocks and time periods come from another way of dividing the rock layers. This way was to divide geological time into four periods – primary, secondary, tertiary and quaternary. Table 12.1 shows the names of the rock layers and the time periods they represent, their meanings and the time when each layer began to form.

1 Can you devise a way to help you remember the order of the time periods? Here is a start you may like to use, or make up your own from the beginning: Queens Take Crowns Just To Please …

2 Which period do we live in and how long has this period existed?

3 The Age of Dinosaurs began about 251 mya and ended about 65 mya. What are the names of the time periods in which dinosaurs lived?

Table 12.1 Rock layers, time periods and time scale

Name of rock layer and time period	Meaning	Time when the layer began to form/ millions of years ago (mya)
Quaternary	the fourth period	about 2.5
Tertiary	the third period	about 65
Cretaceous	named after chalk rocks found in France	about 145
Jurassic	named after the Jura Mountains, found in France, Germany and Switzerland	about 200
Triassic	named after three distinctive layers of rocks found in Germany and Western Europe	about 251
Permian	named after a region in Russia	about 300
Carboniferous	named after the large amounts of coal found in the layer	about 360
Devonian	named after a region of England, called Devon	about 416
Silurian	named after an ancient tribe in Wales, in the United Kingdom	about 444
Ordovician	named after an ancient tribe in Wales	about 488
Cambrian	an ancient name for Wales	about 542

Figure 12.2 These hills in central Queensland, Australia, are made from Cretaceous rocks in which dinosaur fossils have been found.

The rock layers are not just found in the places from which they take their names, but are found all around the world.

In the middle of the 19th century, a Canadian geologist named Sir William Logan (1798–1875) discovered rocks in the Laurentian mountains of Canada that formed before those in the Cambrian layer. These rocks are called Precambrian rocks. Their discovery meant that the Earth was older than the Cambrian period suggested.

How fossils form in rocks

Most **fossils** form from plant and animal bodies. The plant or animal must be covered quickly after death by sediments such as mud and sand. The covering stops **scavengers** ripping up animal bodies. The covering also contains little air and oxygen so decomposers cannot thrive and rot down the bodies. In time, the sediments harden to form rock and the bodies inside them form fossils. They can do this in two ways.

In one process, called replacement, the tissues of the dead body are dissolved and washed away by water passing through the rock. A cavity forms and minerals in the water passing through it come out of the solution and form a solid, which makes a rocky shape of the body.

In the second process, called petrification, water containing dissolved minerals seeps into the tissues, and minerals come out of the solution and strengthen the tissues into rock.

4 If a dinosaur died on a rocky surface, what are its chances of forming a fossil? Explain your answer.

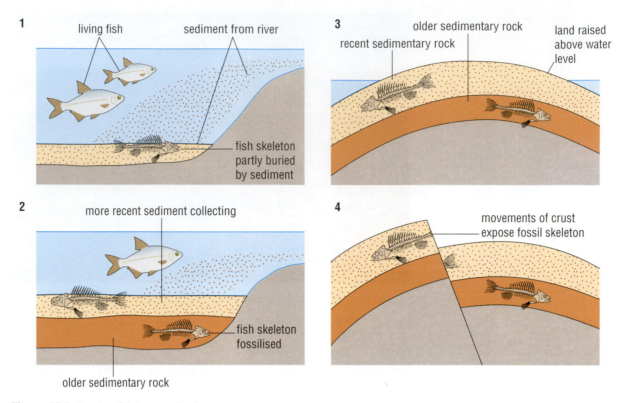

Figure 12.3 How two fish become fossils

Some fossils are not formed by the actual plant or animal body but by things they have left behind such as eggs, solid wastes (their droppings) and tracks. These fossils are known as trace fossils.

How ideas about fossils changed

Fossils have been known since ancient times and there have been many ideas about them. Some were thought to come from the heads of toads, one type was thought to be the horn of a unicorn, while others were thought to be thunderbolts made in storms and then flung into the ground.

Figure B These fossils are ammonites. They were first thought to be the curled-up bodies of stone snakes. Later they were found to be the shells of creatures related to the octopus, which swam in ancient seas.

The word 'fossil' was first used by Georgius Agricola (1494–1555) who was a German doctor. He used the word to describe anything that was dug out of the ground. He included ancient pottery in his definition of a fossil. Later the word fossil was used to describe any stony animal-shaped object. Some people believed that it was just a coincidence that the stones looked like the bodies of animals.

Some triangular-shaped stones had been widely known as tongue stones. Some people believed they grew in the rocks and others believed they fell from the Moon. Nicolaus Steno (1638–1686), a Danish geologist, remembered tongue stones when he was examining the teeth of a shark and noticed how they were similar. This observation led him to believe that the tongue stones were the teeth of ancient sharks, which had been left behind in the rock after the shark had died. He then looked at other fossils and decided that they too had come from the bodies of ancient animals.

3 How did the meaning of the word 'fossil' change in time?

4 **a)** Who made the link between a stone object in the ground and a living thing?

b) How was the link made?

c) Which scientific enquiry skill did he use to make the link?

Fossils and rocks

Once scientists had a clear idea that fossils were the remains of plants and animals that lived in the distant past, they could then start to examine them thoroughly. They found that some fossils were only found in one or a few layers while a few were found in many layers. They found that some were only found in a few places around the world while others were widespread. They also found that some were easy to recognise while others needed close and thorough examination to identify them.

From all this work, geologists discovered that some fossils were found in just one layer. This meant that the organism that formed them lived in just one time period. They also discovered that some of these fossils were found in many parts of the world and they were easy to recognise. This made them very useful in helping to indicate the age of a layer of rock and they are known as indicator or **index fossils**.

Index fossils are particularly useful where sedimentary rocks have formed in the same time period but in different ways. For example, one may be a rock formed from mud while another may be limestone. If the same index fossil is found in both types of rock, geologists can be certain that they formed at the same time and belong to the same time period.

5 What are three characteristics of index fossils?

Table 12.2 Index fossils for different time periods. The international scientific name of the species (page 112) is given together with the common general name of the animal group to which the species belongs.

Time period	Index fossil	Time period	Index fossil
Quaternary	*Pecten gibbus* (scallop)	Permian	*Leptodus americanus* (brachiopod)
Tertiary	*Calyptraphorus velatus* (sea snail)	Carboniferous	*Lophophyllidium proliferum* (coral)
Cretaceous	*Scaphites hippocrepis* (ammonite)	Devonian	*Mucrospirifer mucronatus* (brachiopod)
Jurassic	*Perisphinctes tiziani* (ammonite)	Silurian	*Cystiphyllum niagarense* (coral)
Triassic	*Tropites subbullatus* (ammonite)	Ordovician	*Bathyurus extans* (trilobite)
		Cambrian	*Paradoxides pinus* (trilobite)

6 Which time period is identified by a scallop?
7 Which time periods have ammonites as their index fossils?
8 If you found a rock layer with a brachiopod in it, what index fossil would you look for in the layer beneath?

The fossil record

Each group of living things has a fossil record. This shows when members of the group existed and left fossils. Some groups of living things, like algae and bacteria, have a long fossil record and are found in rocks from the Precambrian onwards. The mammal group has a much shorter fossil record. Mammal fossils are only found in the Jurassic period onwards. Figure 12.4 shows the fossil record of the major groups of living things.

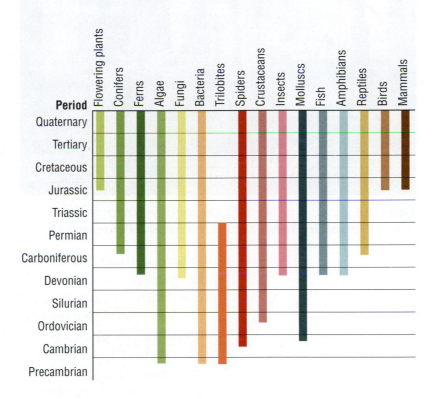

Figure 12.4 The fossil record of living things

9 Which groups of living things existed before the Cambrian period?

10 When did fungi first appear in the fossil record?

11 How many groups of living things were in the fossil record in the Ordovician?

12 Which appeared in the fossil record first – ferns or conifers?

13 How many new groups were added to the fossil record in the Devonian?

14 In which period did the number of groups of living things decrease?

15 Which group has become extinct, and when was it last recorded in the fossil record?

16 In the period during which the flowering plants appeared, which animal groups also appeared in the fossil record?

The fossil record and the age of the Earth

The fossils found in a particular layer can be used to find out about the conditions on Earth at that time.

Figure 12.5 This picture has been made after examining fossils from the Carboniferous period and imagining how the living things may have looked in their environment.

For example, fossils of plants and animals in Carboniferous times are of living things that lived in damp conditions and show that many areas had swamps. The fossils of living things from Permian times show where there were deserts. So the fossils show ancient habitats and how conditions, particularly climate, have changed in different parts of the world. As climate change usually takes a long time, the fossil record hints that the age of the Earth may be great.

Looking at fossils from different layers shows that some of the features in any group of living things have changed over time. Many scientists think that this happens in the following way. A change in a feature may occur in one individual in a species and then be passed on through the generations that follow. In some circumstances, the individuals with the new feature may eventually form a new species. Over a long period of time, this species might even replace the old species, which then might become extinct. This process of species change is called **evolution**.

The fossil record does not give a time when scientists think the Earth formed but it does suggest that the Earth is very old.

17 How do fossils indicate that the Earth might be very old?

Figure 12.6 These fossil skeletons are from different species of horse and come from different layers of rock. Species A was found in a deeper layer of rock than species B.

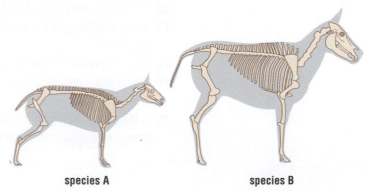

species A

species B

Finding the ages of the rocks

Varves

The first method used to tell the actual age of rocks was developed by the Swedish geologist Gerard De Geer (1858–1943). He studied the building up of sediments in lakes that received water from melting glaciers. The glaciers have a period of melting every year during the summer months, and the water from them carries sand and silt down streams and rivers into lakes. These fragments form a light-coloured band in the sediment. Then, as winter conditions set in, the amount of water entering the lakes slows down and carries only smaller clay particles. These form a darker band above the lighter band of sand and silt. The two bands for each year produce a distinctive layer. These are known as **varves**, from the Swedish word meaning 'layer'. By digging into the sediment and counting the varves, it has been found that the earliest layer was laid down 13 300 years ago. Varves then can be used to measure over 13 000 years back in time.

Radioactive materials

In Chapter 8, page 125, we learnt that the word 'atom' means 'indivisible', but in fact the atoms of some elements do break down and form other atoms. An atom that breaks down in this way is said to **decay**. When an atom decays, it releases a form of energy known as **radiation**. These elements are known as **radioactive elements**. When they have decayed they produce substances that are no longer radioactive. Here is a very simple explanation of how scientists find the age of rocks.

Many rocks contain a radioactive substance, A. As substance A decays, it produces a non-radioactive substance, B. Every million years, half of substance A decays into B. Figure 15.8 shows how the amounts of the two substances in the rock change over time.

Figure 12.7 These varves were laid down in a lake in Washington, USA.

18 Estimate how many years of laying down sediments are shown in the rock you can see in Figure 12.7.

For discussion

Why are varves not used everywhere to count back in time to the beginning of the Earth?

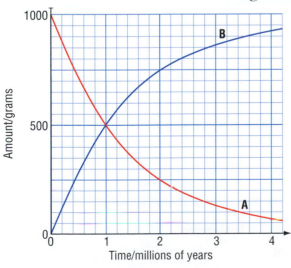

Figure 12.8 How a radioactive substance (A) changes over time into a non-radioactive substance (B)

19 a) How long does it take for 1000g of substance A to decay to $\frac{1}{4}$ of this amount?

b) When 1000g of substance A has decayed for 3 million years, how much of it is still substance A? How much is now substance B?

c) A scientist estimates that a rock is 4 million years old. What would be the amounts of substances A and B in the rock if the estimate is correct?

When scientists look at a rock to find its age they compare the amount of substance A with the amount of substance B. For example, if they find that the rock has equal amounts of A and B (50% of each) they can see that the rock is 1 million years old.

By using radioactive elements in this way, the ages of the layers of rock have been estimated and used to make a geological timescale, as you can see in the right-hand column of Table 12.1. Precambrian rocks have been tested in the same way and the oldest rocks that have been found are 4.6 billion years old. Scientists believe that this is the age of the Earth.

◆ SUMMARY ◆

◆ Rocks form layers (*see page 181*).
◆ Rocks layers have names (*see page 182*).
◆ The layers of rocks and the time periods are arranged into a geological timescale (*see page 182*).
◆ Fossils form in rocks (*see page 183*).
◆ Fossils can be used to identify rock layers (*see page 185*).
◆ Each group of living things has a fossil record (*see page 187*).
◆ The fossil record suggests that the Earth may be very old (*see page 188*).
◆ Varves and radioactive materials can be used to find the age of rocks (*see page 189*).

End of chapter questions

Answer these questions using Table 12.1 and Figure 12.4.

1 Which groups of living things were present in the fossil record a billion years ago?

2 How long have there been flowering plants on the planet?

3 Which groups of living things appeared in the fossil record about 542 million years ago?

4 How long after ferns appeared did conifers appear?

5 Which groups of animals appeared about 200 million years ago?

6 In which time period did one group of invertebrates and two groups of vertebrates appear in the fossil record?

7 What were the groups that appeared in the fossil record in that time period?

8 How long ago did these groups appear in the fossil record?

9 Which group of animals disappeared from the fossil record and when?

10 What happened to this group of animals?

PHYSICS

(13) Measurements

◆ Phenomena and illusions
◆ Measuring length, mass and time
◆ Estimating quantities
◆ Accuracy of measurements
◆ Heat and temperature

Physics is the scientific study of how matter and energy interact. These interactions can produce the colours of the rainbow in a shower or the roar of the wind in a hurricane. At a greater distance, the interactions of matter and energy in the Sun produce light and heat, while inside our eyes light energy is converted into electrical energy, which passes to our brain and allows us to see.

Figure 13.1 The white light in sunbeams is split by the water in rain droplets to produce an arch of coloured bands in the sky.

Every event in the universe, from your next breath to a star exploding, is an interaction of **matter** and **energy**, so physics is really a part of all the other scientific subjects rather than a separate one.

All the information we gather with our senses, such as the presence of light, and events, such as the formation of a rainbow, are called **phenomena** (*singular*: phenomenon). So physics can also be described as the science of investigating phenomena.

Launching a rocket

Sometimes we can observe several phenomena in an event, as the following example of launching a rocket shows.

Figure 13.2 Three, two, one lift off! A multi-stage rocket leaving the launch pad

For discussion

After reading about the rocket launch, a person asked, 'Why was light from the rocket seen before the sound of the rocket was heard? Why did the stages fall back to Earth when they separated from the rocket? Why did the stages burn up in the atmosphere?'

What explanations can you give to answer these questions?

Light from the rocket engines can be seen immediately by the distant spectators as the rocket begins to rise from the launch pad. When the roar of the rocket engines reaches the spectators it nearly deafens them. The rocket's speed increases every second as it rises into the sky.

The rocket is divided into parts, called stages. Each stage has fuel tanks and rocket engines. When the fuel is used up in one stage, that stage will separate from the rocket and fall back towards Earth. As the stage rushes back through the atmosphere it will become so hot that it will burn up. When the last stage has separated, only a small spacecraft will remain in orbit round the Earth or set off across the Solar System.

Fooling our senses

Occasionally what we detect with our senses fools us into thinking we are seeing something else. When this happens we are fooled by an optical illusion.

All our senses can be fooled into seeming to detect something that is not really present. This means that, instead of relying on our senses, we need to make more accurate observations of phenomena in science. We do this by taking measurements.

Length, mass and time

Three things that are measured in many investigations are length, mass and time. Any measurement is made in **units** – for example, a common unit of length is the centimetre. There is an international system of units that is used by scientists throughout the world. This is known as the Système International d'Unités. The units in this system are known as SI units.

Measuring length

The standard SI unit of length is the metre. Its symbol is m, and, as with all symbols for SI units, no full stop is placed after it. The metre is divided into smaller units for measuring small lengths or distances, and large numbers of metres are made into bigger units to measure long lengths or distances. Table 13.1 shows some of these other SI units.

Table 13.1 Units of length

Unit	Symbol	Number of metres
kilometre	km	1000 m
metre	m	1 m
centimetre	cm	0.01 m
millimetre	mm	0.001 m
micrometre	μm	0.000 001 m
nanometre	nm	0.000 000 001 m

Measuring mass

The standard SI unit of mass is the kilogram, whose symbol is kg. The other SI units of mass used in investigations are shown in Table 13.2.

3 What is the mass of the object on each of the balances shown in Figure 13.4? Which was the easiest to read?

Table 13.2 Units of mass

Unit	Symbol	Number of kilograms
megatonne	Mt	1 000 000 000 kg
tonne	t	1000 kg
kilogram	kg	1 kg
gram	g	0.001 kg
milligram	mg	0.000 001 kg

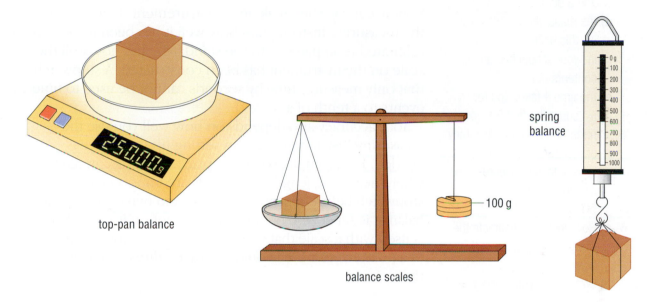

top-pan balance

balance scales

spring balance

100 g

Figure 13.4 Measuring mass

Measuring time

The standard SI unit of time is the second, and its symbol is s. Other units of time used in investigations are shown in Table 13.3.

4 If you saw someone commit a crime, how might you describe to a detective the appearance of the criminal and what happened? Do mass, length and time feature in your answer? If they do, say where they occur.

Table 13.3 Units of time

Unit	Symbol	Number of seconds
day	d	86 400 (or 1440 minutes, or 24 hours)
hour	h	3600 (or 60 minutes)
minute	min	60
second	s	1
millisecond	ms	0.001

Estimating quantities

At the beginning of an investigation, it may be useful to
estimate the quantities that are going to be used or the
time that is going to be taken for certain observations.
At this stage of the investigation, accuracy is not essential
– that comes later.

Accuracy of measurements

Your accuracy when making a measurement depends on
the measuring instrument – how well it has been made and
calibrated (compared with the standard), and how well the
scale on the instrument has been constructed. A stopwatch
that only measures time by seconds cannot be used to time
events to a tenth of a second.

Your accuracy also depends on how well you use the
measuring instrument. Care in setting up the device is
needed. For example, you must place a ruler accurately
when measuring length (both ends are important), reset a
stopwatch before repeating a timing and make sure that a
balance is set at zero before a mass is put on it. If a balance
is used with a scale that is read by looking at the position
of a pointer, your eye should be placed directly in front of
the pointer.

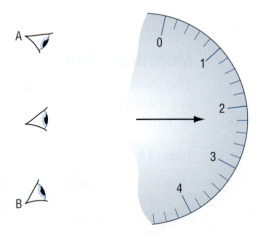

Figure 13.5 It is important to position your eye correctly, looking
horizontally, when taking readings.

Heat and temperature

The hotness or coldness of a substance is measured by taking its temperature. The temperature of a substance is measured on a scale that has two fixed points. The most widely used temperature scale is the Celsius scale. Its two fixed points are 0 °C (the melting point of ice or freezing point of water) and 100 °C (the boiling point of water). In between the two fixed points the scale is divided into 100 units or degrees. The scale may be extended below 0 °C and above 100 °C; laboratory thermometers usually have a scale reading from −10 °C to 110 °C.

A thermometer compares the temperature of the substance in which the bulb is immersed with the freezing point and boiling point of water. It compares the hotness or coldness of a substance.

The lowest possible temperature, known as absolute zero, is −273 °C. Temperatures can go as high as millions of degrees Celsius.

8 How much hotter is:
   a) 45 °C than 30 °C
   b) 20 °C than −15 °C?
9 Why are two fixed points needed for a temperature scale and not just one?

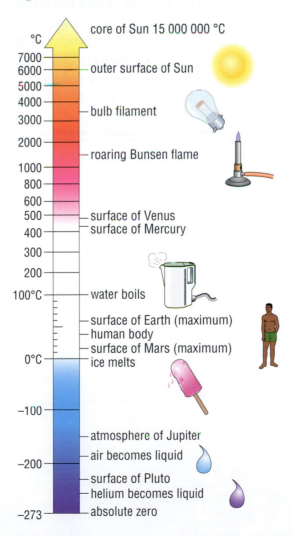

Figure 13.6 The Celsius scale of temperature

Liquids in thermometers

10 Which thermometer, one
containing mercury or one
containing alcohol, could be
used in a polar region where
the temperature reaches below
−40 °C? Explain your answer.

11 Which type of thermometer
could be used to measure the
boiling point of water?
Explain your answer.

Two liquids that are commonly used in thermometers are mercury and alcohol. Mercury has a freezing point of −39 °C and a boiling point of 360 °C. Alcohol has a freezing point of −112 °C and a boiling point of 78 °C.

If the bulb of the thermometer is placed in a hot substance, the liquid inside expands and spreads up inside the thermometer tube. The level it reaches depends on the hotness of the substance, and the temperature can be read from the scale. If the bulb is then placed in a cold substance, the liquid inside contracts back down the tube. The level at which it settles depends on the coldness, and again the temperature is read from the scale.

◆ SUMMARY ◆

- ◆ Phenomena are investigated in physics (*see page 192*).
- ◆ Our senses can fool us into making incorrect observations (*see page 194*).
- ◆ Investigations are carried out by measuring length, mass and time (*see page 194*).
- ◆ It is useful to estimate quantities at the beginning of an investigation (*see page 196*).
- ◆ There are techniques for accurate measuring (*see page 196*).
- ◆ Temperature is a measure of the hotness or coldness of a substance (*see page 197*).
- ◆ The expansion and contraction of liquids in thermometers are used to measure temperature (*see page 198*).

End of chapter questions

1 A model rocket was launched and measurements of its height and horizontal distance from the launch pad were taken as shown in the table.

a) Plot the flight path of the model rocket from these measurements.

b) What was the vertical distance from the launch pad when the rocket stopped rising?

c) What was the horizontal distance from the launch pad when the rocket stopped rising?

d) What was the horizontal distance from the launch pad when the rocket hit the ground?

e) If the rocket had been 10 m above the ground when its horizontal distance from the launch pad was 2 m, where do you think the rocket would have landed?

Horizontal distance from launch pad/m	Vertical distance from launch pad/m
0	0
1	4
2	8
3	11
4	13
5	14.2
6	15
7	15.5
8	15
9	13
10	10
11	0

2 A volume of liquid was heated and cooled. During this time the temperature of the liquid was measured nine times. The results are shown in the table.

a) Plot a graph of the data in the table.

b) What do you think the temperature was at:
　i) 5 minutes
　ii) 7 minutes?

c) Did the liquid heat up and cool down at the same rate? Explain your answer.

d) What would you estimate the temperature to be after 13 minutes?

Time/min	Temperature/°C
0	20
1	40
2	60
3	80
4	90
6	96
8	80
10	60
12	40

♦ What forces do
♦ Different types of forces
♦ Friction
♦ Air resistance
♦ Mass and weight
♦ Gravity and weight
♦ How springs stretch

Forces and their effects

1 Describe all the pushing and pulling forces shown in Figure 14.1.

You cannot see a force but you can see what it does. You can also feel the effect of a force on your body. A **force** is a push or a pull.

Forces can:

- make an object move – for example, if you throw a basketball your muscles exert a pushing force on the ball and it moves through the air when you let it go
- make a moving object stop – for example, a goalkeeper moves into the path of a moving ball to exert a pushing force on the ball to stop it
- change the speed of a moving object – for example, a hockey player uses a hockey stick to push a slow-moving ball and send it shooting past a defender
- change the direction of a moving object – for example, a batsman can change the direction of a cricket ball moving towards the wicket by deflecting it so that it moves away from the wicket towards the boundary
- change the shape of an object – for example, when a racket strikes a tennis ball, part of the ball is flattened before the ball leaves the racket.

Figure 14.1 Forces act in many ways.

Figure 14.2 In cricket, forces make the ball move, stop, change direction, change speed and even change shape.

How to measure a force

A force can be measured with a newton spring balance. The SI unit for measuring force is the newton (symbol N). This force is quite small and is equal to the gravitational force on (the **weight** of) an average-sized apple, or the pulling force needed to peel a banana!

Different types of forces

There are two main types of forces – contact forces and non-contact forces. A contact force occurs when the object or material exerting the force touches the object or material on which the force acts. A non-contact force occurs when the objects or materials do not touch each other.

Contact forces

All the situations described so far are examples of contact forces in action. Some more examples follow.

Impact force

When a moving object collides with a stationary object an impact force is exerted by one object on the other.

The size of the force may be large, such as when a hammer hits a nail, or it may be very tiny, such as when a moving molecule of gas in the air strikes the skin.

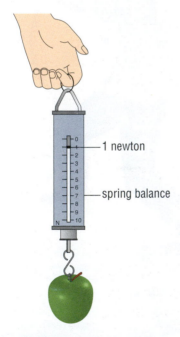

Figure 14.3 The apple pulls on the spring balance with a force equal to its weight.

1 newton

spring balance

Figure 14.4 Impact forces can also occur between two moving objects, as can be seen in this stock car race. This motorsport is popular in Brazil, Canada, Great Britain, New Zealand and the USA.

Strain force

When some materials are squashed, stretched, twisted or bent, they exert a force that acts in the opposite direction to the force acting on them. These materials are called elastic materials and the force they exert when they are deformed is called a strain force. When the force applied to the material is removed, the strain force exerted by the material restores the deformed material to its original shape. For example, the strain force in a squashed tennis ball as it is hit returns the ball to its original shape when the ball has left the racket.

Tension is a strain force that is exerted by a stretched spring, rope or string. At each end the tension force acts in the opposite direction to the pulling force.

2 **a)** i) What do you feel if you hook the two ends of an elastic band over your index fingers and slowly move your hands apart?
ii) What happens to the elastic band when you bring your hands together again?
iii) What happens if one end of a stretched elastic band is released?
b) Describe how the strain force changes in each part of question **a)**.

A force is shown in a diagram as an arrow pointing in the direction of the push or the pull.

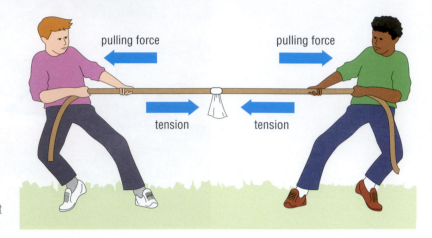

Figure 14.5 Tension is a strain force that acts against the force applied.

Friction

Friction is a contact force that occurs between two objects when there is a push or a pull on one of the objects that could make it move over the surface of the other object. Friction acts to oppose that movement.

As the push or pull on the object increases, the force of friction between the surfaces of the objects also increases. This force matches the strength of the push or the pull up to a certain value and so, below this value, the object does not move. The friction that exists between the two objects when there is no movement is called static friction.

If the strength of the push or pull on the object is increased beyond this value, the object will start to slide. There is still a frictional force between the two surfaces, acting on each surface in the opposite direction to the direction of its movement. This frictional force is called sliding friction. The strength of this force is less than the maximum value of the static frictional force.

Figure 14.6 Friction between the log and the ground opposes the pulling force of the horse.

3 Imagine you are asked to push a heavy box across the floor. At first you need to push very hard but once the box has started to move you can push less strongly yet still keep it moving. Why is this?

4 When you take a step forwards you push backwards on the ground with your foot. Make a sketch of Figure 14.7 and draw in an arrow to show the frictional force that stops your foot slipping.

5 Figure 14.8 shows a wheel turning. Make a sketch of the wheel. Draw and label:
a) the force exerted by the wheel pushing backwards on the road
b) the force of friction preventing the wheel slipping.

Figure 14.7

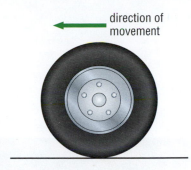

direction of movement

Figure 14.8

A closer look at friction

The surfaces of objects in contact are not completely smooth. Under a microscope it can be seen that they have tiny projections with hollows between them (Figure 14.9).

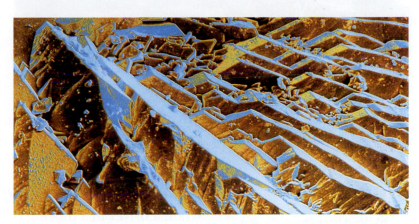

Figure 14.9 A metal surface that appears smooth to the naked eye has projections, which can be seen when it is magnified 180 times.

Where the projections from the surface of one object meet the projections from the surface of the other, the materials in the projections stick. These connections between the surfaces produce the force of friction between the objects.

Reducing friction

If a liquid is placed between the two surfaces, the projections are forced apart a little and the number of connections is reduced, which in turn reduces the force of friction. This can cause problems or it can be helpful. For example, water running between the surface of a tyre and the road reduces the friction between them and increases the chance of skidding. However, oil between the moving metal parts of an engine and the parts in the bearings reduces friction and also reduces wear on the metal parts.

6 Why does oiling the axles of a bicycle make the bicycle move more easily?

7 How could you find out if cooking oil reduces friction more than water?

Increasing friction

The friction between two surfaces can be increased by pressing the surfaces together more strongly.
This makes the projections press against each other more, and increases the size and number of connections between the surfaces.

Figure 14.10 Tyres with treads are designed so that water squirts out from between the treads.

When brakes are applied on a bicycle or car, the brake pads press against a moving part of the wheel and the force of friction increases. This opposes the rotation of the wheel and slows down the bicycle or car until it stops.

The tread on a car tyre is designed to move water out of the way as the tyre rolls over a wet road, reducing the risk of skidding. Racing cars have smooth tyres that are ideal for a dry track. If it rains, they slide and skid all over the place and the tyres need to be changed.

Friction and road safety

When a driver in a moving car sees a hazard ahead, the car travels a certain distance before the driver reacts and applies the brakes. The distance travelled by the car in this time is called the thinking distance. This is followed by the braking distance, which is the distance covered by the car after the brakes are applied and before the car stops. Table 14.1 shows the thinking and braking distances that will bring a car with good brakes to a halt on a dry road.

Table 14.1

Speed	Thinking distance/m	Braking distance/m	Total stopping distance/m
48 km/h (30 mph)	9	14	23
80 km/h (50 mph)	15	38	53
112 km/h (70 mph)	21	75	96

8 a) What happens at the beginning of the time during which a car covers the thinking distance?

b) What happens at the end of the time during which the car covers the thinking distance?

9 What may affect the thinking time of the driver? How would the thinking distance of the car be affected? Explain your answer.

10 What, other than speed, may affect the braking distance of the car? Explain your answer.

11 A car is travelling along a road at 80 km/h when a tree falls across the road 54 m away. What would probably happen and why?

For discussion

How safe is:

a) driving close to the car in front

b) driving fast on winding country roads with high hedges?

Explain your answers to each part of the question.

'It's the driver that's dangerous, not the car.'
Assess the usefulness of this slogan for a road safety campaign.

Other forces affecting speed

When objects move along a surface, friction occurs and opposes the motion. Two other forces that affect speed are air resistance and water resistance.

Air resistance

Air is a mixture of gases. When an object moves through the air it pushes the air out of the way and the air moves over the object's sides and pushes back on the object. This push on the object is called air resistance or drag.

The value of the air resistance depends on the size and shape of the object. Many cars are designed so that the air resistance is low when the car moves forwards. The car's body is designed like a wedge to cut its way through the air and the surfaces are curved to allow the air to flow over the sides with the minimum drag. Shapes that are designed to reduce air resistance are called **streamlined** shapes.

A dragster is a vehicle that accelerates very quickly. In a dragster race, two vehicles accelerate along a straight track. At the end of the race the dragsters are slowed down by brakes and a parachute. The parachute offers a large surface area against which the air pushes. The high air resistance of the parachute slows down the dragster and helps it stop in a short distance.

Figure 14.11 Testing a streamlined car in a wind tunnel

The air resistance produced by a parachute is also used to bring sky divers safely to the ground. The resistance of the gases in the atmospheres of other planets in the Solar System is used to slow down space probes so they can land safely and the devices on board are able to carry out their investigations.

12 How would the size of parachute required on a space probe to allow it to land safely differ on:
 a) a planet such as Venus which has a thick atmosphere
 b) a planet such as Mars which has a thin atmosphere?
Explain your answers.

Figure 14.12 A safe landing for the Mars Exploration Rover

Water resistance

When an object moves through water it pushes the water out of the way, and the water moves over the object's sides and pushes back on the object. This push on the object is called water resistance or **drag**. Objects that can move through the water quickly have a streamlined shape. A fish such as a barracuda, which moves quickly through the water, has a much more streamlined shape than a slow-moving sunfish.

Figure 14.13 A shoal of barracuda (left) and a sunfish (right)

Water resistance affects the movement of ships and boats on the water surface. Boats designed for high speeds have a hull shaped to reduce water resistance as much as possible. Some boats are equipped with a device called a hydrofoil, which reduces the area of contact between the boat and the water so that water resistance is kept to a minimum. The boat (itself called a hydrofoil) can then move quickly over the water surface.

Figure 14.14 A hydrofoil

N = north pole
S = south pole

Figure 14.15 Bar, horseshoe and ring magnets

13 If you had a magnet with its north and south poles marked on it and a magnet without its poles marked, how could you identify the poles of the unmarked magnet?
Explain your answer.

Non-contact forces

Non-contact forces include magnetic forces, electrostatic forces and gravitational forces. They all exert their force without having to touch the object.

Magnetic force

A magnet has a north-seeking pole and a south-seeking pole. These are usually known as the north pole and the south pole. If you pick up two magnets and bring either their north poles or their south poles together, you will feel a force pushing your hands apart as the two similar poles repel each other. You will feel your hands being pushed away even though the magnets are not touching. The strength of the push increases as you bring the two similar poles closer together.

If you bring the north pole of one magnet towards the south pole of another magnet, you will feel your hands being pulled together as the different poles attract each other. The strength of this pull increases as the poles get closer together.

A magnet can also exert a non-contact force on objects made of iron, steel, cobalt or nickel. Either pole of the magnet exerts a pulling force on these magnetic materials. The strength of the force increases as the magnet and the magnetic material are brought closer together.

Figure 14.16 This 'Maglev' train is supported above its track by strong magnetic forces. It travels quietly on a 'cushion' of air which eliminates friction between the train and the tracks.

Gravitational force

There is a force between any two masses in the universe. The masses may be small, such as those of an ant and a pebble, or they may be very large, such as those of the Sun and the Earth. The force that exists between any two masses because of their mass is called the gravitational force. The force acting between small masses is too weak to have any noticeable effect on them but the gravitational force between two large masses such as the Sun and the Earth is large enough to be very important. It is the gravitational force between the Sun and all the planets in the Solar System that holds the planets in their orbits (Figure 11.3, page 159). The gravitational force between an object on the Earth and the Earth itself pulls the object down towards the centre of the Earth and is called the weight of the object.

Mass and weight

The **mass** of an object is a measure of the amount of matter in it. The **weight** of an object is the pull of the Earth's gravity on the object. For example, an object may have a mass of 1 kg. The pull of the Earth's gravity on 1 kg is a force of almost 10 newtons (actually 9.8 N but it is often rounded up to make the calculations easier). The weight of the 1 kg mass is therefore 10 N.

The region in which a force acts is called a **field**. There is a gravitational field around the Earth. The gravitational field strength is calculated by the equation:

$$\text{gravitational field strength} = \frac{\text{weight}}{\text{mass}}$$

At the Earth's surface we have seen that the pull on a mass of 1 kg is 10 N, so the gravitational field strength is 10 N/kg.

14 What is the weight of each of the following masses on Earth?
a) 2 kg
b) 3.5 kg
c) 5.25 kg

15 What is the weight of a 6 kg object on the surface of Mars?

16 It is planned to bring samples of Mars rock back to the Earth. If 50 kg samples were collected by a space probe robot, what would be the weight of each sample on:
a) Mars
b) Earth?

17 The Moon's gravitational field strength is one-sixth that of the Earth. What would be the weight of a 1 kg object on the Moon?

18 A sample of Moon rock weighed 30 N on the Moon.
a) What would be its weight when it was brought to the Earth?
b) What is the mass of the sample?

The gravitational field strength on the surface of Mars is three times less than the gravitational field strength on the surface of the Earth. This means that a 1 kg object that is part of a space probe would have a weight of 10 N when it was on Earth but a weight of only 3.3 N on the surface of Mars.

The mass of an object remains the same wherever it goes in the universe but its weight changes according to the gravitational force that is acting upon it.

Weightlessness

The gravitational field strength around a planet, moon or star gets weaker and weaker as you move further away.

A space station in orbit above the Earth is still in the Earth's gravitational field. The force of gravity pulls on the space station but because the space station is moving with a velocity parallel to the surface of the planet, the force pulls the space station so that its path is curved. The force is just enough to keep the space station at exactly the same height. It does not move closer to Earth but 'falls' in a circular path around it.

Inside the spacecraft every object that is not held down floats about. These objects, including the astronauts, are also 'falling' around the planet in the same way as the space station. The floating state is called apparent weightlessness because it feels like having no weight but the objects are, in fact, still being pulled by the Earth's gravity.

You may feel something similar to this weightlessness for a moment when you begin moving downwards in an elevator or travel on a ride at a fair where you fall directly downwards. Both you and the ride are falling, so you briefly feel lighter than usual. You may feel heavier just as the ride stops.

True weightlessness could only occur far out in deep space where there are no large objects with gravitational fields. This is beyond the distance travelled by any space exploration undertaken so far.

Figure 14.17 These people are enjoying a fairground ride where motion affects their apparent weight.

19 The gravitational field strength on the Moon is one-sixth that of the Earth. If you lived on the Moon how would the weight of some everyday objects such as your watch, school bag or sports shoes change?

20 What would be the weights of the everyday things you considered in question **19** on:
 a) Mercury
 b) Jupiter?

Gravity and weight

We have all known about the force of gravity, from our earliest days, as the force that makes things fall. Although we may think of gravity as the force that pulls things down to the ground, it does more than that. It pulls them towards the centre of the Earth.

Objects released by an astronaut on the Moon fell to the Moon's surface. This indicates that the Moon also has a gravitational force. However, if an object is weighed on Earth and then weighed again on the Moon, its weight will be seen to have decreased. The decrease in weight is due to the less powerful gravitational force on the Moon. This is due to the mass of the Moon being much less than the mass of the Earth. The gravitational force of an object depends upon its mass.

Table 14.2 shows how the force of gravity exerted at the surfaces of different planets compare. The Earth's gravitational field strength is taken as 1 – the standard by which the others are compared.

Table 14.2 The planets' gravitational field strengths compared

Planet	Gravitational field strength
Mercury	0.38
Venus	0.9
Earth	1
Mars	0.30
Jupiter	2.64
Saturn	0.925
Uranus	0.79
Neptune	1.12

How springs stretch

Robert Hooke (1635–1703) investigated the way in which springs stretched when masses were attached to them. He first hung up a spring and measured its length without any mass attached to it. He then hung a mass on the bottom and measured the new length of the spring. He calculated the extension of the spring by subtracting its original length from the new length with the mass attached. Hooke repeated the experiment with different masses. Each time he found the total extension by subtracting the original length from the new length. He found that as the mass increased, the size of the extension increased in proportion – the extension of the spring was proportional to the mass attached to it.

21 A spring is 6 cm long when it is unstretched but is stretched to 9 cm when a mass is hung from it. What is the extension of the spring?

Each time Hooke removed the mass the spring returned to its original length. However, he eventually placed a mass on the spring that stretched the spring so much that it remained slightly stretched when the mass was removed. The spring had gone beyond a point called the **elastic limit** and was permanently deformed. When a larger mass was then added to the spring it no longer extended in proportion to the mass. The spring beyond its elastic limit was in a state known as plastic deformation.

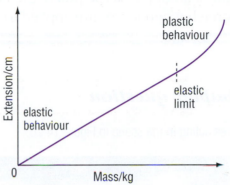

Figure 14.18 How the extension of a spring varies with the mass attached to it

22 An unstretched spring is 6 cm long but becomes 7 cm long when a 100 g mass is hung from it. The spring becomes 8 cm long when a 200 g mass is hung from it.
   a) What is the extension for each mass?

   b) What extension do you predict when masses of
   i) 300 g
   ii) 350 g
   are hung from it in turn? Can you be sure that the extension values you predict will in fact occur? (Hint: think about the elastic limit.)

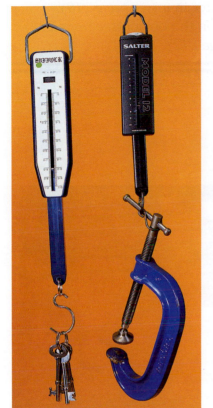

Figure 14.19 Spring balances with different scales

23 How do you think the spring in a spring balance with a scale of 0–10 N compares with the spring in a spring balance that measures forces up to 500 N?
24 A spring balance without a stop would not give correct readings for the weights of the masses hung from it if large masses were used. Explain the reason for this.

The newton spring balance

The discovery made by Robert Hooke has led to the development of a force measurer using a spring that is not stretched beyond its elastic limit. This instrument is called a spring balance. The extension of the spring, and hence the reading on a scale, is proportional to the weight of the mass hung from it, or the force with which it is pulled. The scale of the balance is calibrated in newtons so it is sometimes called a newton spring balance or a newtonmeter.

There is a range of spring balances that measure forces of different sizes. For example, a spring balance may measure forces with values in the range 0–10 N, 0–100 N or 0–200 N.

There is a device called a stop on most spring balances. It prevents the spring from stretching beyond its elastic limit.

◆ SUMMARY ◆

◆ A force is a push or a pull (*see page 200*).

◆ Friction is a contact force that acts to oppose movement (*see page 204*).

◆ The push of the air on a moving object is called air resistance (*see page 207*).

◆ The push of water on an object moving through water is called water resistance (*see page 208*).

◆ The mass of an object is a measure of the amount of matter in it (*see page 210*).

◆ The weight of an object is the pull of the Earth's gravity on the object (*see page 212*).

◆ Springs stretch when forces are applied to them (*see page 212*).

End of chapter question

Identify the forces acting in the scene in Figure 14.20.

Figure 14.20

◆ What is energy?
◆ Forms of energy
◆ Energy changes
◆ Wasted energy
◆ Fuels

What is energy?

In everyday language, we use the word 'energy' in many different ways. Just look at the examples in Figure 15.1.

Figure 15.1 Different ways of using the word 'energy'

The scientific way of thinking about energy is that it is the property of something that makes it able to exert a force and do work. To understand this, it is helpful to think about the ways that energy is stored and what happens when it is changed from one form to another.

Forms of energy

There are two kinds of energy – stored energy and movement energy. Stored energy is also called **potential energy** because it gives something the potential to use its stored energy, as we shall see in the examples. Movement energy is also called **kinetic energy**. The word 'kinetic' comes from a Greek word meaning 'motion'. There are several forms of each kind of energy.

Gravitational potential energy

The force of **gravity** between an object and the Earth pulls the object towards the centre of the planet. If an object is in a position above the surface of the Earth, it possesses stored energy called gravitational potential energy.

Examples of objects with this type of stored energy are plates on a table, books on a shelf, a child at the top of a slide and an apple growing on a branch. Each of these objects is supported by something but if the support is removed they will accelerate to the Earth's surface and their potential energy will be released and changed into other forms.

1 If you are holding this book, or if it is resting on a table or desk, why does it possess potential energy?

2 If you held a stone over the mouth of a well and then let it go, what would happen to the stone? Explain your answer.

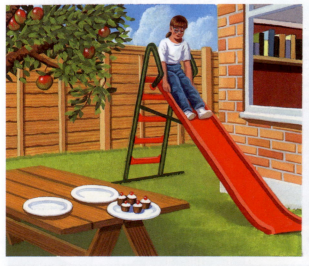 

Figure 15.2 When the objects fall, their stored potential energy is released.

Strain energy

Strain energy is also called elastic potential energy. Some materials can be easily squashed, stretched or bent, but spring back into shape once the force acting on them is removed. They are called **elastic materials**. When their shape is changed by squashing, stretching or bending they store energy, which will allow them to return to their original shape.

A spring stores energy when it is stretched or squashed. Gases store strain energy in them when they are squashed. For example, when the gas used in an aerosol is squashed into a can it stores strain energy. Some of this is used up when the nozzle is pressed down and some of the gas is released in the spray.

3 Look at Figure 15.3. When is elastic potential energy stored and when is it released in:
   a) a toy glider launcher
   b) the elastic cords or springs beneath a sun-lounger
   c) a diving board?

4 Does stretching an elastic band further store up more energy? Plan an investigation to answer this question.

Figure 15.3 Places where strain energy can be stored and released

Chemical energy

Energy can be stored in the chemicals from which a material is made. The chemicals are made from **atoms** that are linked together to make **molecules.** The chemical energy is stored in the links between the atoms. Food, fuel and the chemicals in an electrical cell (or battery) are examples containing stored chemical energy.

The energy is released when the links between some of the atoms are broken and the molecule in which the energy was stored is broken down into smaller molecules. For example, carbohydrates are a store of chemical energy in food. During **respiration**, carbohydrate is broken down into carbon dioxide and water. The energy that is released in this process is used by your body to keep you alive. The energy released by a fuel is used to heat homes, to heat water to produce steam for generating electricity in power stations, and for the production of new materials.

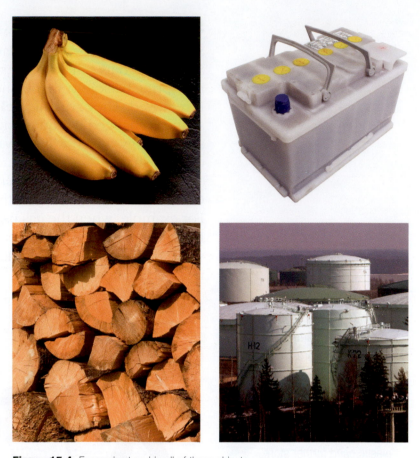

Figure 15.4 Energy is stored in all of these objects.

Kinetic energy

Any moving object has kinetic energy. The object may be as large as a planet or as small as an atom and because of its motion it can do work. When an object with kinetic energy strikes another object, a force acts on them both that will distort the second object or set it moving. For example, if you move your foot and kick a stationary ball, the ball moves away.

Sound energy

Sound energy is produced by the vibration of an object such as the twang of a guitar string. The energy passes though the air by the movement of the atoms and molecules. They move backwards and forwards in an orderly way. This makes a wave that spreads out in all directions from the point of the vibration. Sound energy can also pass through solids, liquids and other gases. The atoms move in a similar way to the turns on a slinky spring when a 'push–pull' wave moves along it.

5 Look out of a window and make a list of everything you can see that has kinetic energy.

Figure 15.5 The kinetic energy of the demolition ball is transferred to the building and breaks up its structure.

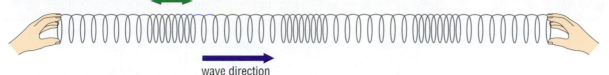

to-and-fro vibration of the turns as the push-pull wave passes

wave direction

Figure 15.6 A slinky spring shows how sound waves move.

Electrical energy

Electric current is the movement of electric charges through a conductor such as copper or graphite. The electric charges are given electrical energy by the battery and carry it to the working parts of a circuit. This may be a lamp, for example, where the energy is changed into light and heat.

> **6** In what ways is electrical energy put to work in your home?

Internal energy

Internal energy is also called thermal energy. All substances are made up of particles. They possess a certain amount of energy, which allows them to move. When a substance is heated this movement increases. For example, the particles in a solid are moving backwards and forwards about a fixed position. The particles in a liquid move more quickly and can move past each other. The particles in a gas can move freely in all directions at high speeds. When a substance is heated, the particles receive more energy and move faster.

Electromagnetic energy

There is a form of energy that can travel through space at the speed of light. This kind of energy travels in waves that have some properties of electricity and some properties of magnetism. They are called **electromagnetic waves**. As these waves make up rays of light and heat, this form of energy is sometimes called radiation energy.

There is a huge range of possible wave sizes, or wavelengths. A wavelength is shown in Figure 15.7.

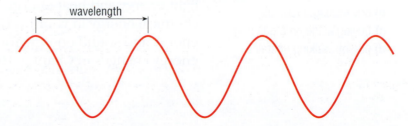

wavelength

Figure 15.7 A wave showing wavelength

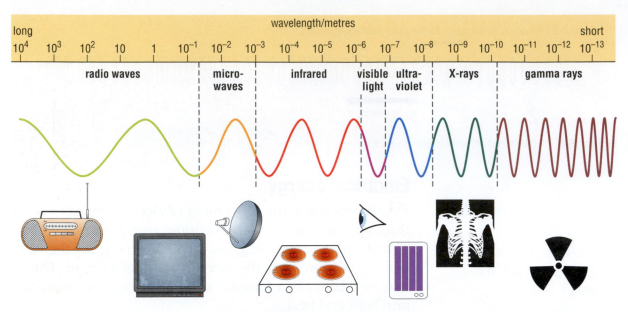

Figure 15.8 The electromagnetic spectrum

Electromagnetic waves are split into seven groups according to wavelength, as Figure 15.8 shows. The different groups have different properties and different uses. The two most familiar groups are light and radio waves.

Light energy

Light is the energy that we detect with our eyes. The light energy escaping from the Sun can be spread out by a prism or a shower of raindrops into light of different wavelengths. This forms the colours of the rainbow (Figure 13.1, page 192) because our eyes see different wavelengths of light as different colours.

Energy changes

We use energy in many ways – for example, to cook food, light our homes and move cars and buses. When energy is used it always changes from one form to another and some always changes into heat energy. For example, when you switch on a light, electrical energy is changed into light energy and heat energy. When you play a guitar, chemical energy in your body is changed into movement energy and sound energy. You can find out more about energy changes in Chapter 16.

7 Which radiation energy has:
a) the longest waves
b) the shortest waves?
8 Which radiation energy can our eyes detect?

9 Say what main energy change takes place in the following examples.
a) clockwork toy
b) boy kicking a football
c) boiling kettle on a gas ring
d) person walking upstairs

Figure 15.9 Energy changes occur when a guitar is played.

Wasted energy

When we turn on a lamp, it is because the light is useful to us. We do not use the heat that is produced so it is wasted. Sometimes that wasted energy can cause problems.

For example, some machines make so much noise (wasted sound energy) that people using them have to wear ear protection (Figure 15.10).

10 What is the wasted energy in each of the energy transfers in question **9**?

Figure 15.10 These workers' ears are protected from noise energy.

Fuels

Many substances are burned to release their chemical energy to provide heat and light. They are called **fuels**. Wood, coal, gas, charcoal, oil, diesel oil, petrol, natural gas and wax are examples of fuels. The heat may be used

Figure 15.11 In Nepal, ovens are often fuelled by wood.

to warm buildings, cook meals, make chemicals in industry, expand gases in vehicle engines and turn water into steam to generate electricity. Some gases and waxes are used to provide light in homes, caravans and tents.

Coal, gas and oil were all formed from plants and animals that lived millions of years ago, so they are known as **fossil fuels**.

Fossil fuels

Coal is formed from large plants that grew in swamps about 275 million years ago. These plants used energy from sunlight in the same way that plants do today. When they died they fell into the swamps. There was a lack of oxygen in the swamp water, which prevented bacteria growing and decomposing the dead plants. Eventually the plants formed peat. Later the peat became buried and was squashed by the rocks that formed above it. The increase in pressure squeezed the water out of the peat and warmed it. These processes slowly changed the peat into coal.

Tiny plants and animals live in the upper waters of the oceans and form the plankton. When they die, they sink to the ocean floor. Over 200 million years ago, the dead plankton that collected on the ocean floor did not decompose because there was not enough oxygen there to allow bacterial decomposers to live. The remains instead formed a layer, which eventually became covered by rock. The weight of the rock squeezed the layer and heated it. This slowly converted the layer of dead plankton into oil and methane gas. This is the gas that is supplied to homes as natural gas. Several fuels are obtained from oil.

11 What conditions helped fossil fuels to form?

Renewable resources

Unfortunately, the supplies of fossil fuels are limited and there will come a time when there are not enough to meet our needs. As a result, scientists are trying to develop alternative sources of energy from renewable energy sources such as the movement of the wind, the movement of waves and the tide, the movement of water from rivers (hydroelectricity) and the light of the Sun (solar power).

◆ SUMMARY ◆

◆ Energy is the property of something that makes it able to exert a force and do work (*see page 216*).

◆ Stored energy is called potential energy (*see page 216*).

◆ Movement energy is known as kinetic energy (*see page 216*).

◆ Chemical energy is energy that is stored in the chemicals from which a material is made (*see page 217*).

◆ Sound energy is transferred by waves in which atoms move backwards and forwards (*see page 218*).

◆ A battery gives electrical energy to electric charges, which allows them to move through a conductor as an electric current (*see page 219*).

◆ When a substance is heated, the particles it is made from move faster (*see page 219*).

◆ Electromagnetic energy is transferred by electromagnetic waves (*see page 219*).

◆ Energy can change from one form to another (*see page 220*).

◆ Some energy is wasted when it changes (*see page 221*).

◆ Substances that are burnt to release their chemical energy are called fuels (*see page 221*).

End of chapter questions

1 A group of students was investigating the potential energy in a nail 15 cm long. They suspended it above a block of soft clay, measured the distance to its tip, and then let it go. The students measured the depth to which the nail sank in the clay. The table shows their results for four experiments.

a) How do you think they measured the depth of the indent in the clay?

b) Plot a graph of their results.

c) How could you use the graph to predict the indent made by the nail from a height greater than 1 m?

Height of nail above clay/cm	Depth of indent/cm
25	0.9
50	1.6
75	2.3
100	3.0

2 A second group of students investigated the potential energy of a brass sphere, which was dropped from different heights into soft clay. They measured the diameter of the indent made by the sphere. The table shows their results for four experiments.

a) How do you think the students measured the diameter of the indent?

b) Plot a graph of their results.

c) How do these results compare with the results of the experiment described in question **1**?

d) Suggest a reason for any differences you describe.

e) Can the graph be used to predict indentations produced by falls from any height greater than 70 cm? Explain your answer.

Height of sphere above clay/cm	Diameter of indent/cm
5	1.0
20	1.7
50	2.2
70	2.5

16 Energy transfers

◆ How energy use has increased
◆ Measuring work
◆ Energy transfer diagrams
◆ Sankey diagrams
◆ Plants and energy
◆ Energy and ourselves
◆ Generating electricity
◆ Conservation of energy

Energy transfers and transformations

You are living at this moment because of energy transfers taking place in your body. As the energy is transferred it is transformed. Stored chemical energy in your food is transformed to kinetic (movement) energy when you raise your hand to turn the pages of this book. Some of the stored chemical energy is also transformed into heat to keep your body warm.

All the changes you can detect around you are due to energy transformations. If you are reading this book by an electric light, electrical energy is being transformed into light energy so that you can see the words. If you can hear someone shuffling about in their seat next to you, some of the kinetic energy of their body is being transformed into sound energy that reaches your ears. Also at each energy change or transformation some energy is always lost as heat energy.

1 If you are wearing a watch, what energy changes are taking place in it right now?

How energy use has increased

Energy transformations are vital to survival. From the earliest times people have needed enough stored chemical energy in their bodies to change to kinetic energy to move around and find food. If they could not find enough food (and store it as chemical energy for later use), they simply starved to death. In time, people came to live together in groups and developed machines and other devices to make life easier.

A city in the past

Figure 16.1 A city scene from about 120 years ago

Figure 16.1 shows a busy square in Manchester, England, about 120 years ago. The insides of buildings were lit by candles, oil lamps and gas lights. Coal and wood were used as fuel for fires to keep the buildings warm in winter. The streets were lit by gas lights. People and goods were transported through the streets by carriages and carts pulled by horses. Many people arrived or left the city by trains pulled by steam locomotives. Some people walked.

A city today

Figure 16.2 A modern city

Figure 16.2 shows Times Square in New York, today. The buildings are lit by electricity. Some buildings have air conditioning powered by electricity to keep them cool in summer. They also have heaters powered by electricity to keep them warm in winter. Cars, vans, trucks and buses are used to transport people and goods. These vehicles are powered by petrol or diesel engines and are also used by people travelling to and from the city. Many people may also use trains pulled by a locomotive with an electric motor. Some people may come to the city by aircraft, which use a fuel called kerosene, similar to petrol.

2 How have the sources of energy for light and movement in cities changed over the last 120 years?

3 Draw energy transfer diagrams for:
   a) winding up a clockwork car
   b) letting a clockwork car run
   c) letting a battery-powered car run.

4 Diagram D shows the energy transfer for a camera using a photographic film. What would the energy transfer diagram be for taking a picture with a digital camera in a mobile phone?

5 Diagrams A, C and D show the main energy changes. However, there should be a second energy output for complete accuracy. What is this output? Explain your answer. (You may like to turn to page 221 to help you answer.)

Energy transfer diagrams

Energy transformations can be shown by energy transfer diagrams. There are three parts to an energy transfer diagram:

1 an arrow showing the energy input
2 a box showing an energy converter or transducer
3 arrows showing energy output.

Here are some examples of energy transfer diagrams.

A Releasing a catapult

strain energy → ⟨catapult⟩ → kinetic energy

B Burning gas in a Bunsen burner

stored chemical energy → ⟨Bunsen burner⟩ → light energy and heat energy

C Blowing up a balloon

kinetic energy → ⟨balloon⟩ → strain energy

D Taking a photograph

light energy → ⟨camera⟩ → chemical energy

Figure 16.3 Releasing a catapult

Figure 16.4 Blowing up a balloon

Sankey diagrams

A Sankey diagram is a second kind of diagram that shows energy changes. It features arrows of different widths. The width of the arrow indicates the amount of energy it represents. The unit in which energy and work are measured is called the joule (symbol J). A thousand joules is a kilojoule and is represented by the symbol kJ.

Energy in a car engine

A car engine has an energy input of 200 kJ. The energy output is 80 kJ of kinetic energy and 120 kJ of heat energy.

6 What investigation could you make with a balloon to find out if different amounts of stored strain energy change into different amounts of kinetic or sound energy when the energy is transferred?

7 Draw a Sankey diagram for a light bulb that has an energy input of 100 J and an energy output of 10 J of light energy and 90 J of heat energy.

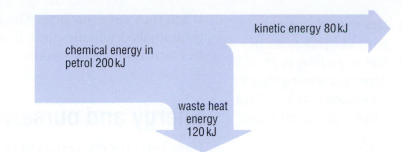

Figure 16.5 This Sankey diagram shows what happens to the energy passing through a car engine every second.

Energy from the Sun

Figure 16.6 is a diagram showing the path of energy reaching the Earth from the Sun.

Plants and energy

When a seed germinates, its skin breaks open and the tip of the root pops out (Figure 19, page 18). Energy stored inside the seed is used as the root grows and seeks out water.

Stored energy is also used by the growing shoot. The shoot grows up through the soil and eventually reaches the surface. Energy from below, in the seed, is used as the shoot sends out leaves.

Some of the light energy falling on the leaves is trapped inside them. It is converted into stored chemical energy as the plant makes food using water from the soil and carbon dioxide from the air. This process of making food using light energy is called **photosynthesis**.

The chemical energy stored in a plant is transferred to a herbivorous animal when it eats it. The herbivorous animal then has a store of chemical energy, which it uses to keep itself alive and to move about. Carnivorous animals feed on herbivorous animals. They take in stored chemical energy

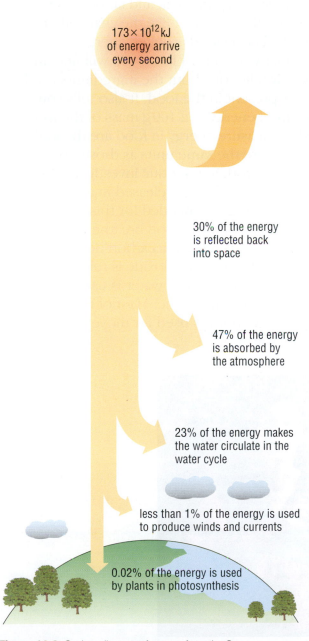

Figure 16.6 Sankey diagram of energy from the Sun

8 Draw an energy transfer
diagram for photosynthesis.

9 What do you think might
happen to a seedling shoot if it
was growing from a seed that
had been planted too deeply?

when they eat their prey. Thus the energy stored and used by animals originally came from light energy from the Sun that was trapped by plants.

Energy and ourselves

Have you ever cooked meat on a barbecue? If you have, you might have seen fat dropping through the grill and bursting into flames as it hit the hot charcoal below. If you did not take enough care when cooking the meat, you may even have seen it catch fire too. The meat and fat burn because they contain chemical energy, just like the charcoal fuel that is heating them. In fact, all foods contain energy.

If you look at many packets of food, you will find an information box. It tells you about the ingredients used and the nutrients present in the food. It also tells you about the amount of energy in a 100 g mass of the food. The units used to measure energy in food are the joule and kilojoule. They are the same units as those used to measure energy and work in scientific investigations.

The chemical energy in food is released in a process called **respiration**. Oxygen is needed for this process and your body takes it in from the air you breathe into your lungs. When the energy is released, carbon dioxide and water are produced. The carbon dioxide is released into the air when you breathe out. The water is used in your body or released in sweat and urine. Most of the energy that is released in your body is used for movement and for keeping the body warm.

Nutrition

Typical Composition	This pack (450g) provides	100g (3¹/₂oz) provide
Energy	2610kJ	580kJ
	621kcal	138kcal
Protein	13.2g	2.9g
Carbohydrate	82.3g	18.3g
of which sugars	18.0g	4.0g
Fat	26.6g	5.9g
of which saturates	13.5g	3.0g
mono-unsaturates	10.4g	2.3g
polyunsaturates	2.7g	0.6g
Fibre	7.2g	1.6g
Sodium	1.8g	0.4g

A serving (450g) contains the equivalent of approx. 4.5g of salt.

Figure 16.7 Food packet label

10 Draw an energy transfer
diagram for the body. The
energy input is the stored
chemical energy in food.

Figure 16.8 This runner is releasing a lot of energy.

Generating electricity

Electrical energy is a very useful form of energy because it is easy to generate and can be transported quickly to wherever it is needed. These properties make it the most widely used form of energy in the highly developed countries of the world.

Michael Faraday (1791–1867), an English physicist, discovered that an electric current could be made to flow in a wire if the wire was made to move through a **magnetic field**. This principle is used to generate electricity in a bicycle dynamo and in a power station generator.

The bicycle dynamo

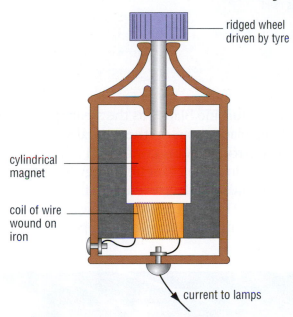

ridged wheel driven by tyre

cylindrical magnet

coil of wire wound on iron

current to lamps

Figure 16.9 Inside a bottle or sidewall dynamo

11 What would be the problem if dynamos were the only way electricity was provided in bicycle lamps?

A bicycle dynamo is an electrical device which can be located in the hub of a wheel (hub dynamo) or clamped onto the bicycle frame close to a tyre (bottle or sidewall dynamo). The bottle dynamo has a wheel on top which can be made to touch the tyre. The inside of a bottle dynamo is shown in Figure 16.9.

When the dynamo wheel is in contact with the tyre it rotates as the bicycle wheel turns. Inside the dynamo the magnet turns and its field sweeps through the wires, generating an electric current that lights the bicycle's lamps.

The power station generator

Inside a power station, there is a generator consisting of a huge electromagnet surrounded by coils of wire. The electromagnet is attached to a shaft to which turbine blades are attached (Figure 16.11). When the turbines are made to spin, the electromagnet also spins, generating an electric current in the surrounding coils of wire.

In about two-thirds of the world's power stations, water is heated to make steam. This takes place in a boiler. The energy that the water molecules receive increases their kinetic energy so much that they move apart from each other to form a gas – steam. The steam expands rapidly and exerts a force, which drives it from the boiler to the turbine blades. Here as much as possible of the steam's kinetic

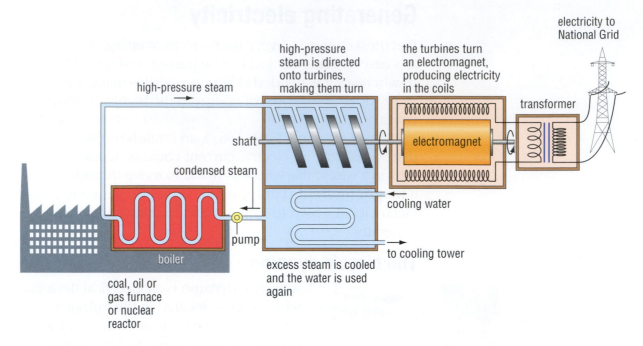

high-pressure steam

high-pressure steam is directed onto turbines, making them turn

the turbines turn an electromagnet, producing electricity in the coils

electricity to National Grid

transformer

shaft

electromagnet

condensed steam

cooling water

pump

to cooling tower

boiler

excess steam is cooled and the water is used again

coal, oil or gas furnace or nuclear reactor

Figure 16.10 The parts of a power station

energy is passed to the turbine blades as the steam pushes past them, making the blades spin on the central shaft.

The generator's electromagnet is connected to the end of the shaft. As it spins using kinetic energy from the turbine blades it generates an electric current in the coils of wires surrounding it. The electricity flows away from the power station to towns and cities in overhead power lines or underground cables.

12 Draw an energy transfer diagram:
a) for the power station boiler
b) for the spinning electromagnet.
13 Describe how life in your home would change if none of the electrical appliances worked.

For discussion

How would your daily life have to change if electricity was no longer generated at any power stations in your country?

Figure 16.11 This turbine assembly for a power station in England is like many others around the world.

Conservation of energy

Imagine that you have a piece of string with a small weight on the end. It can be used as a pendulum. If you held it up in front of you and pulled the weight towards your face until it almost touched your nose, the pendulum would be ready to swing. You might think that when you let go of the weight it would swing away from you and then back again, and perhaps hit your nose. If you were to try this experiment you would find that your nose was safe. The pendulum would not reach your nose when it swung back. In fact, as the weight swung to and fro it would approach your nose less and less, until it stopped.

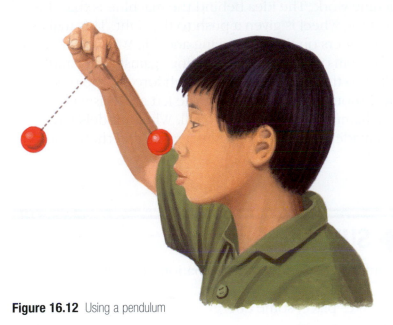

Figure 16.12 Using a pendulum

It would seem that as the pendulum swung back the first time some energy was lost. It is true that pendulums do lose energy as they swing but the energy is not destroyed. As the string and weight move through the air, they rub against the air particles. This rubbing causes heat energy to be released from them, just like the heat energy that is released from your hands when you rub them together. With each swing of the pendulum, more heat energy is lost until all the energy has left the pendulum and it hangs vertically, motionless.

This experiment shows that energy is not destroyed, it just changes form. Just as energy is not destroyed, so it is not created either. The energy used in the body or a machine simply comes from somewhere else in the universe. This discovery about energy led to the Law of Conservation of Energy which says that energy cannot be made or destroyed, it can only be changed from one form to another.

Some inventors have tried to defy the Law of Conservation of Energy by designing and even making perpetual motion machines. Figure 16.13 shows an example.

14 What is the energy transfer diagram for the pendulum when it:
   a) is released close to someone's nose
   b) stops moving at the other end of its swing?

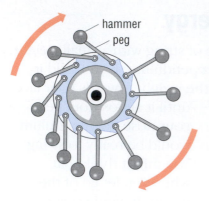

Figure 16.13 Perpetual motion machine

15 Why will the perpetual motion machine shown in the picture not run forever?

This perpetual motion machine is called an overbalanced wheel. It is believed that someone in India thought of the idea for this machine about 1200 years ago. Since then many people in other parts of the world have worked on the idea but all have failed to make the machine work. The idea behind the machine is that the top of the wheel is given a push to the right. This makes the hammers on the right swing and fall. When they reach the bottom of the wheel they swing against a peg and push it to the left. If there is enough force in the push, the wheel should turn upwards on the left and this will cause more hammers on the right to fall. When models have been made, they work for a short while and then stop.

◆ SUMMARY ◆

◆ An energy is transferred, it is tranformed. Energy transformations are energy changes (*see page 224*).

◆ When an energy transformation takes place, some energy is always lost as heat (thermal) energy (*see page 224*).

◆ Energy transfers can be shown by energy transfer diagrams, which have three parts – energy input, energy converter and energy output (*see page 226*).

◆ Work and energy are measured in a unit called the joule (*see page 226*).

◆ Energy transfers can be shown by Sankey diagrams with arrows of different widths (*see page 226*).

◆ Seeds contain stored energy, which is used for germination and early growth of seedlings (*see page 227*).

◆ Some light energy falling on the leaves of a plant is converted into stored chemical energy inside the plant (*see page 227*).

◆ Chemical energy in food is released in respiration (*see page 228*).

◆ Electrical energy is easy to generate and transport (*see page 229*).

◆ A bicycle dynamo generates electricity (*see page 229*).

◆ Energy cannot be made or destroyed (*see page 231*).

End of chapter question

On your way home from school, make a list of examples of energy transformations.
When you get home, draw an energy transformation diagram for each example.

◆ Movements in the sky (and how the Earth moves)
◆ Lights in the sky
◆ Measuring with light
◆ The Moon
◆ Early studies of the Solar System
◆ Discovering Uranus and Neptune
◆ The parts of the Solar System
◆ The Milky Way galaxy
◆ Planets around other stars
◆ Beyond the Milky Way galaxy

The Earth formed from rocks and dust moving round the Sun about five billion years ago, as described in Chapter 11. In this chapter, we are going to examine the Earth in space.

Movements in the sky

If you were to watch the sky for a day and a night, you would see the Sun rise towards the east at dawn. It would continue to rise in the sky until midday and then it would sink in the sky and set towards the west. As the sunlight faded, the sky would darken and other stars would be seen to cross the sky from east to west before they faded as the sunlight appeared again in the sky.

1 Draw a horizontal line and write 'east' at the left end and 'west' at the right end. Now draw in the path of the Sun across the sky.

Figure 17.1 The Sun shines constantly in space as the Earth rotates on its axis.

People once believed that the stars really were moving across the sky, but today we understand that the Sun and the stars do not change position in this way. It is the daily rotation of the Earth that makes them appear to move.

The **axis** about which the Earth rotates is not perpendicular to the plane of the Earth's orbit. If it were, the Sun would rise to the same height in the sky each day of the year. The axis is at an angle of about 23° to the perpendicular and remains pointing in the same direction throughout the Earth's orbit (Figure 17.2).

Figure 17.2 The changing seasons in each hemisphere as the Earth progresses in its orbit

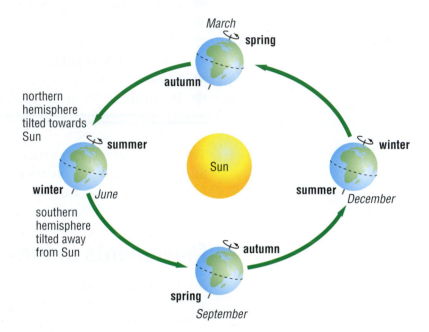

We divide the Earth into two half spheres or hemispheres. They meet at the equator, which is an imaginary line running around the middle of the planet between the poles. The hemispheres are known as the northern hemisphere and the southern hemisphere.

As the Earth moves in its orbit there is a time of year when the northern hemisphere is tilting towards the Sun and the southern hemisphere is tilting away from it. Six months later the northern hemisphere is tilting away from the Sun and the southern hemisphere is tilting towards it.

These changes in the way each hemisphere tilts towards and away from the Sun can cause changes in the length of day and night, and in the strength of the sunlight reaching an area of the Earth's surface. This produces the periods of time called seasons.

The east-to-west path of the Sun across the sky changes with the position of the Earth in its orbit. When a hemisphere is tilting towards the Sun, the path of the Sun

2 What is the position of the Earth (in which direction is the axis tilted) when it is summer in:
   a) the northern hemisphere
   b) the southern hemisphere?
3 What is the position of the Earth when the Sun rises to its lowest midday position in the sky in:
   a) the northern hemisphere
   b) the southern hemisphere?
4 What is the position of the Earth when:
   a) the day is longer than the night in the southern hemisphere
   b) the day is shorter than the night in the southern hemisphere
   c) the day and the night are the same length of time?

is different from the path when the hemisphere is tilting away from the Sun. Sunrise is earlier, the Sun rises higher in the sky at midday, and it sets later in the evening. Figure 17.3 shows the path of the Sun across the sky when the hemisphere is tilted towards the Sun (mid-summer), away from the Sun (mid-winter) and when it is changing from tilting in one direction to the other (at the spring and autumn **equinoxes**).

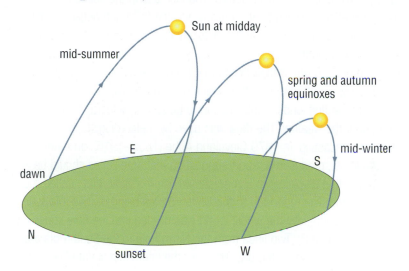

Figure 17.3 The changing path of the Sun across the sky as the seasons change (in the northern hemisphere)

Time zones

Up until the late 19th century, every town kept its own time, called local time. This was based on the position of the Sun as it appeared in the sky and the shadow cast by a sundial. Mechanical clocks were invented in the 1650s and could be adjusted to keep time with the sundial. This allowed time to be measured on cloudy days or at night.

Communication between towns in these times was slow. People walked, rode horses or travelled in horse-drawn vehicles. When people reached the next town they would have to reset their watches to the local time.

In the 1830s, Joseph Henry (1797–1878), an American physicist, invented the telegraph. He discovered that an electrical current sent on a long wire could cause a small electromagnet to move.

Figure A An early telegraph transmitter in use

Henry could control the movement of the distant electromagnet by the way he opened and closed the switch in the circuit. This allowed messages to be sent very quickly between places that were many kilometres apart. Samuel Morse (1791–1872), an American inventor, devised the Morse Code, which was used on the telegraph system.

While work was being done to improve the telegraph system, railways were being set up across the United States and elsewhere. They could carry people faster than ever before. The increase in the speed of communication provided by the railway and the telegraph began to make the use of local times confusing. For example, if you agreed to send a message to a person at 2 pm your local time, it might reach them at 1.50 pm or 2.10 pm at their local time, simply because the local time at one place was different from the local time at other places.

One of the first kinds of information to be sent by telegraph was data about the weather. The data was used by meteorologists in weather forecasting. In 1879 Cleveland Abbe (1838–1916), an American meteorologist, produced a report suggesting that time zones should be set up and that all places in a zone should use the same time, called the standard time. The report led to the United States being divided into four time zones. Eventually the whole world was divided into the time zones we use today (Figure B). Two lines from the North Pole to the South Pole are important in dividing up the day. They are the international date line, which runs to the east of Japan and New Zealand, and the Greenwich meridian, which runs through an area of London in the UK called Greenwich.

1 How could you set your watch to local time using a sundial?
2 How do you think local times would affect people using a train timetable?
3 When it is midday in London, what time is it:
  a) three zones to the east
  b) five zones to the west?
4 When it is 6 am in Riyadh what time is it:
  a) three zones to the east
  b) four zones to the west?

5 Which three pieces of apparatus did Joseph Henry choose to use when he made his discovery about electricity?

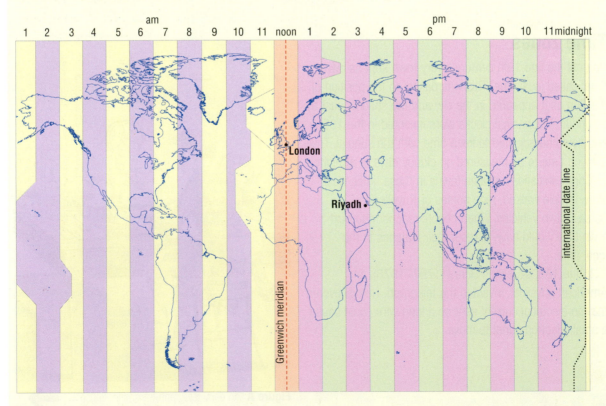

Figure B The world's time zones

The changes in the starry sky

As the Earth turns during the night, stars rise over the eastern horizon and move across the sky in an arc to the western horizon. Stars near either pole appear to move around in a circle during the night. As the Earth moves around in its orbit, it passes different groups of stars and they appear in the night sky. They are known as winter, spring, summer and autumn stars.

Lights in the sky

The Sun is the only large luminous body in our Solar System. We see the other large objects, such as the Moon, the planets and comets, by the sunlight they reflect to the Earth. As all these objects are relatively close to the Sun and the Earth, they reflect light quite strongly towards the Earth. The other stars are luminous but much further away. Although they generate their own light, the beam is very weak by the time it reaches the Earth.

The movement of the gases in the Earth's atmosphere does not significantly affect the strong light beams from the Moon, the planet and comets – they shine steadily in the sky. The weak light beams from the stars, however, are affected and the result is that their light does not shine steadily but appears to flicker. So a planet, such as Venus, shines brightly in the night sky against a background of twinkling stars.

5 How can you tell a planet from a star in the night sky?

Measuring with light

The vast distance between two objects in space can be measured by the time it takes light to travel between them. For example, the time taken for light to travel between the Sun and the Earth is about eight minutes. The time for light to travel between two stars is much longer and is measured in **light years**. A light year is the distance travelled by light in a year. This distance is 9.5 million million kilometres.

The nearest star to the Sun is Proxima Centauri, which is 4.3 light years away. This star and the Sun are just two of the 500 000 million stars in the Milky Way galaxy. This is a group of stars that is 100 000 light years across. There are about 100 000 million other galaxies in the universe. They are great distances from our own. For example, the Andromeda galaxy, which can be seen as a fuzzy patch with the naked eye, is 2.2 million light years away.

Figure 17.4 The Andromeda galaxy seen through a telescope

6 a) Look at Table 14.1.
Which star is the brighter,
Betelgeuse or Spica?

b) Why might you expect Spica
to be above Betelgeuse in
the table?

c) Betelgeuse is a red giant
star. How could this
information help you explain
its position in the table?

Bright stars

The brightness of a star depends on its size, its temperature and its distance from the Earth. Table 17.1 shows some stars arranged in decreasing order of brightness as seen from Earth.

Table 17.1 The features of some bright stars, arranged in decreasing order of brightness

Star	Temperature/°C	Colour	Distance from Earth
Sun	6 000	yellow	8 light minutes
Sirius	11 000	white	8.6 light years
Arcturus	4 000	orange	36 light years
Betelgeuse	3 000	red	520 light years
Spica	25 000	blue	220 light years

Constellations and planets

The stars make patterns in the sky. These patterns are called constellations. The arrangement of the stars in a constellation is due to their position in space, which in turn is due to chance. The stars may seem to be grouped together at the same distance from the Earth but they are not. Some stars in a constellation may be many light years closer to the Earth than others.

While the stars appear to be fixed in their positions, the planets do not. Each night a planet is found in a different position from the previous night. The name 'planet' comes from the Greek word meaning 'wanderer'. The planets seem to wander across the night sky against the background of constellations. This is due to the way the Earth and the planets move around the Sun in their **orbits**.

The Moon

The Moon moves round the Earth in about 28 days. Only the side of the Moon's surface that is facing the Sun reflects light, so as its orbit progresses the illuminated part that we can see from Earth changes shape. The different shapes are known as phases of the Moon (Figure 17.5).

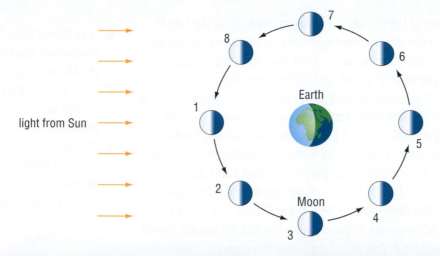

light from Sun →

shape of Moon seen from Earth:

| 1 New Moon | 2 | 3 | 4 | 5 Full Moon | 6 | 7 | 8 |

Figure 17.5 Phases of the Moon

As the Moon moves around the Earth, it also spins on its axis. The speed at which it rotates makes the Moon always keep the same part of its surface facing the Earth. This is why the surface of the Moon always appears the same and the other side of the Moon is never seen from the Earth. It has been seen by astronauts and photographed by space probes in orbit around the Moon.

Early studies of the Solar System

The movements of the Sun, Moon, stars and planets across the sky were studied by ancient civilisations. They used the movements of the Sun and the Moon to measure time. This helped them to plan when to sow seeds to raise crops.

It seemed to the Ancient Greeks that all the objects in the sky were set in crystal spheres, which moved around the Earth. However, one Greek philosopher, called Aristarchos (c.320–250 BCE), suggested that the movements of the planets could be explained by considering them to move around the Sun. The other Greek philosophers were not enthusiastic about this idea and preferred their model featuring crystal spheres.

Figure C The arrangement of the Sun and planets around the Earth according to the Ancient Greeks

The arrangement of the planets in the crystal spheres did not fully fit with the observations made of their movements in the sky. For example, Mercury and Venus did not move so far from the Sun as the crystal sphere arrangement suggested they should, and the apparent backward motion of Mars, Jupiter and Saturn that occasionally occurred could not be explained by this model. The Egyptian astronomer Ptolemy, who lived in the 2nd century, proposed an explanation. If the planets moved around the Earth in a looped circular path (Figure D), the observed planetary motion would result. This explanation satisfied most people for about 1300 years.

Nicolaus Copernicus (1473–1543), a Polish astronomer, challenged these ideas – and the Earth-centred view of the Church – by suggesting that the Sun was at the centre of the universe and that the planets moved in circular orbits around it. This model supported the observed movements of the planets. For example, the apparent backward motion of Mars, Jupiter and Saturn could be explained by the Earth 'overtaking' them as it moved in its orbit around the Sun.

Tycho Brahe (1546–1601) was a Danish astronomer who made detailed observations of the planets and stars before the invention of the telescope. In 1577, a large **comet** appeared in the sky. People believed that comets were high clouds in the atmosphere. Brahe measured the position of the comet as it moved across the sky. He discovered that it was further away than the Moon and not part of the atmosphere. He also discovered that it moved in an elliptical path, passing through space where the crystal spheres were thought to be without being restricted to movement within one.

Like most people, Brahe could not believe in the Sun-centred model of Copernicus, although his observations suggested that the crystal spheres did not exist. He devised a model of his own. In this model he placed the Earth at the centre of the universe as the Greeks had done because he was reluctant to give up the ideas he had been taught.

6 How do you think the ancient civilisations used the Sun and the Moon to measure time?

7 How do you think a knowledge of time helped in the production of crops?

8 Why do you think it made sense to the Ancient Greeks to consider the Earth at the centre of the universe?

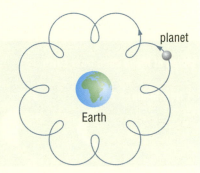

Figure D Planetary motion according to Ptolemy

Figure E Brahe's observatory

9 Brahe collected a huge amount of data on planets, stars and the comet. How did he gather it?

Galileo Galilei (1564–1642) was a professor of mathematics at the University of Padua in Italy. He built a telescope and examined the skies with it. He discovered what appeared to be stars around Jupiter and recorded their positions.

Galileo thought that if these stars were like others in the universe then the movement of Jupiter must be unlike that of the other planets. He decided that Jupiter probably moved just as the other planets did and that the 'stars' were really moons moving around Jupiter, just as the Moon moves around the Earth. This was proof that some things in the sky did not move around the Earth but moved around other objects.

Johannes Kepler (1572–1630), a German astronomer, worked with Tycho Brahe. When Brahe died, Kepler re-examined the vast amount of data Brahe had collected. Kepler studied the orbits of the planets. He found that the data accurately matched orbits of an elliptical shape around the Sun. This discovery provided the final evidence against the Earth-centred universe and established our current knowledge about motion in the Solar System.

At the time that Galileo and Kepler were investigating the Solar System, experiments on magnetism by an English scientist and doctor called William Gilbert (1544–1603) were becoming widely known. Galileo and Kepler thought that a magnetic force might hold the objects in the Solar System in their places, too.

Isaac Newton (1642–1727) is believed to have begun his investigation on the force that holds objects in the Solar System when he saw an apple fall from a tree.

He reasoned that the force that pulled the small apple down to the Earth might also pull the large Moon, even though the Moon was at a much greater distance from the Earth. He considered the Moon to be falling towards the Earth in its orbit around the Earth. The Moon did not reach the Earth, he thought, because it was at such a great distance and the strength of the force was weaker there. This resulted in the Moon falling *around* the Earth rather than *on* to it.

From his calculations on this force of gravity on objects near the Earth, Newton predicted the rate of fall of the Moon needed to give its movement in a curve around the Earth. When Newton made calculations on the actual movement of the Moon he found that they matched his prediction. From this work he showed that the objects in the Solar System moved due to the force of gravity.

The structure of the Solar System and the way things move in it due to gravity were worked out when it was thought that only six planets were present.

10 Which scientific enquiry skill did Kepler use to make his discovery about elliptical orbits?

Figure F Did the same force pull on the Moon?

11 How did the work of Brahe, Galileo and Kepler destroy the idea of crystal spheres around the Earth?

12 Why was it not unreasonable to believe that magnetism might be an important force in holding objects in the Solar System?

13 In what way did chance play a part in the discovery that gravity is the force that acts between objects in the Solar System?

Discovering Uranus and Neptune

William Herschel (1738–1822) was born in Germany but spent most of his life in England. He and his sister Caroline (1750–1848) ground discs of metal into mirrors and made telescopes with them. They made the best telescopes of their time, and used them to examine the sky in detail.

In 1781, William Herschel discovered a spot of light in the sky that did not twinkle like a star. At first he thought it was a comet but he discovered that the edges were not fuzzy like a comet's but clear like those of a planet. More observations were made on the object. It was discovered to move in an orbit that was a similar shape to the orbits of other planets and not like the greatly elongated ellipse of a comet's orbit. Herschel wanted to name the new planet after the King of England, George III, while some astronomers wanted to call it Herschel. Eventually it was decided to follow the tradition of naming planets after ancient gods and call the planet Uranus.

There is a gravitational force that exists between any two objects in the universe. The gravitational force of a planet affects the paths of other planets around the Sun. When the path of Uranus was observed, it seemed as if the planet was being affected by the gravitational force of another planet. Urbain Leverrier (1811–1877), a French astronomer, calculated the position of this other planet and asked Johann Galle (1812–1910), a German astronomer, to look for it. On the first evening of the search in September 1846, Galle located the planet with his telescope. It was named Neptune.

Figure G William Herschel (1738–1822) and Caroline Herschel (1750–1848)

14 What evidence suggested that Herschel had discovered a planet?
15 Who do you think should be credited with the discovery of Neptune?

16 Recognising results and observations that do not fit into a pattern is a scientific enquiry skill. How was this skill important in the discovery of Uranus and Neptune?

The parts of the Solar System

The planets

The part of the Solar System that is most well known is made up of the Sun and eight planets, as shown in Figure 11.3 (page 159). The planets move around the Sun in elliptical orbits in an anticlockwise direction.

7 Arrange the eight planets in Table 17.2 in order of mass, starting with the most massive planet.
8 Arrange the eight planets in order of diameter, starting with the planet with the largest diameter.
9 Does the diameter of a planet give you an indication of its mass?
10 Does the mass of a planet give you an indication of its distance from the Sun, or its rotation time?

Table 17.2 Planet data

Planet	Diameter/km	Mass (Earth = 1)	Distance from Sun/ million km (approx.)	Rotation time			Orbit time/days
				days	hr	min	
Mercury	4878	0.056	58	58	15	30	88
Venus	12100	0.82	108	243	0	0	224
Earth	12756	1	150		23	56	365
Mars	6793	0.107	228		24	37	686
Jupiter	142880	318	778		9	50	4332
Saturn	120000	95	1427		10	14	10759
Uranus	50800	14.5	2871		10	49	30707
Neptune	48600	17	4497	6	15	48	90777

11 The asteroid belt lies between the orbits of two planets. Which ones? Use Table 17.2 to help you.

12 Which asteroids could crash into the Earth?

For discussion

It has been calculated that the asteroid 1950DA will hit the Earth in 2880. What should be done?

Asteroids

Asteroids are lumps of rock that move in orbits around the Sun. They formed in the early stages of the Solar System (see page 159), and range in size from grains of sand to Ceres, the largest asteroid, which is 913 km across. Most of the asteroids move in orbits between 300 and 500 million kilometres from the Sun. They form a huge ring of space rubble called the asteroid belt. Some asteroids have orbits further away from the Sun and a few have orbits that take them across the Earth's orbit.

The Solar System does not end at the orbit of Neptune. Beyond this is a huge ring of rocky and icy objects, which make up the Kuiper Belt. In this belt are found dwarf planets. The largest of them is called Pluto. Some comets

13 Look at Figure 17.6 and estimate the width of the asteroid Ida.

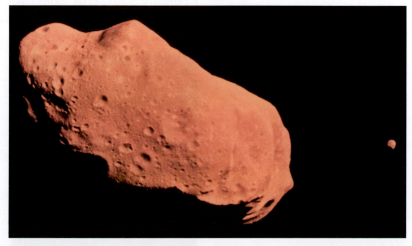

Figure 17.6 This asteroid is called Ida and is nearly 58 km long. It has a moon called Dactyl, which is about 1.5 km across. Can you see it?

are thought to come from the Kuiper Belt while many others are thought to come from the edge of the Solar System, which is called the Oort Cloud. This is believed to be a huge hollow ball of icy objects. If you look at the diagram of the Solar System in Figure 11.3 (page 159), imagine that your head is the Sun, hold out your arms and move them around, your finger tips are in about the region where the Oort Cloud is thought to be.

Figure 17.7 The tail of the comet reflects sunlight which makes it visible in the sky.

The Milky Way galaxy

The work of many astronomers has shown that the Sun is not the centre of the universe as people once believed. It is on an arm of a spiral galaxy called the Milky Way as shown in Figure 17.8. This galaxy received its name from the pale white glow it makes across the sky as shown in Figure 17.9.

Figure 17.8 The Milky Way: top view (left) and side view (right)

Figure 17.9 The Milky Way as seen from Earth

The Milky Way galaxy is 100 000 light years across and rotates like a huge pin wheel in space at 970 000 kilometres per hour. This means that the Sun in its position about 28 light years from the centre of the galaxy takes 225 million years to go round once. The last time the Sun was in its current position, dinosaurs were just starting to develop on the Earth!

Planets around other stars

The power of telescopes has greatly increased since the planets Uranus and Neptune were discovered, and more improvements are being made every year. In 1988, some astronomers believed they had found a planet moving around a star 45 million light years away and over the years, as telescopes continued to improve, over 400 more planets were discovered. Many are gas giants like Jupiter but in 2010 a space telescope called Kepler – in orbit above the Earth where the gases in the atmosphere do not interfere with starlight – began detecting many more. Some of them may be planets similar to Earth.

When scientists look for stars that might have planets, they look for two things. First they look to see if the star wobbles. If it does, this means that there is a force of gravity between the star and one or more planets around it and this force is making the star move. Second they look to see if the star dims occasionally and in a regular pattern. If it does, it means that a planet is crossing between the star and the telescope and reducing the amount of light reaching the telescope. Once a star that wobbles and dims has been identified, the scientists can look closer to find out more about the objects in orbit around it.

For discussion

Some stars are believed to have planets. What do you think are the chances of simple life forms existing on a planet somewhere in the universe?

How would you rate the chances of discovering each of the following?
a) simple life forms
b) intelligent life forms with a lower technological development than us
c) intelligent life forms with a higher technological development than us
Explain your answers.

How will the vast distances between planets in different parts of the universe hinder communications if we find intelligent life?

Beyond the Milky Way galaxy

As early telescopes improved, astronomers began to see smudges of light in the Milky Way galaxy. They discovered that they were clouds of gas in which stars formed.

In the first half of the 20th century, an American astronomer called Ernest Hubble (1889–1953) used the largest telescope then invented to discover that some of the gas clouds were in fact vast groups of stars way outside the Milky Way. These vast groups are separate galaxies and today astronomers estimate that there are billions of them in the universe and that each one may have over a billion stars.

Figure 17.10 Galaxies of stars as revealed by the Hubble space telescope, launched in 1990, which was named after the discoverer of galaxies beyond the Milky Way.

There are three types of galaxies – spiral galaxies like the Milky Way and Andromeda (see Figure 17.4), elliptical galaxies and irregular galaxies, which do not have a definite shape.

Galaxies are moving through space and sometimes they crash into each other. When this happens, the shapes of the galaxies change or they may merge to form a new one. If one galaxy is large and the other is small, the small one may be torn apart as the two crash together. Small galaxies are thought to have crashed into the Milky Way galaxy in the past and some scientists think there may be a chance that the Andromeda galaxy will crash into the Milky Way in a few billion years from now.

◆ SUMMARY ◆

◆ The changes of day and night and the seasons are due to the rotation (spin) of the Earth, the tilt of its spin axis and its orbital path around the Sun (*see page 233*).

◆ The Sun and stars are luminous objects (*see page 237*).

◆ The Moon, planet and comets reflect the Sun's light (*see page 237*).

◆ The phases of the Moon are due to the Moon's orbit around the Earth (*see page 238*).

◆ The ideas and discoveries of Copernicus, Brahe, Galileo, Kepler, Newton, Herschel and Galle have had a great impact on our knowledge of the Solar System (*see page 239*).

◆ The planets are at different distances from the Sun and take different times to move around it (*see page 242*).

◆ Our Solar System is part of a spiral galaxy called the Milky Way galaxy (*see page 244*).

◆ Astronomers have discovered planets around other stars (*see page 245*).

◆ Beyond the Milky Way galaxy, there are billions of other galaxies in the universe (*see page 246*).

End of chapter questions

There is a unit of measurement called the astronomical unit. It is the average distance of the Earth from the Sun. Here are the average distances from the Sun of the first six planets in the Solar System in astronomical units:

   1.0   0.4   9.6   5.2   0.7   1.52

1 Produce a table of these data, using Figure 11.3 (page 159) or Table 17.2 to name the planets.

Scientists look for patterns in their data. In the 18th century, astronomers looked for a pattern in the orbital distances in the following way.

Start with the number sequence 0, 3, 6, 12 … Then continue to double the last number.

Add 4 to each number, and then divide the total by 10.

For example:

   $0 + 4 = 4$ $4 \div 10 = 0.4$
   $3 + 4 = 7$ $7 \div 10 = 0.7$

The resulting numbers are called Bode numbers after the German astronomer Johann Bode (1747–1826).

2 Calculate the first seven Bode numbers in order and write down how they compare with the orbital distances of the planets. What conclusions can you draw from these values?

3 When Uranus was discovered it was found to have a distance from the Sun of 19.2 astronomical units. How did this fit in with the relationship known as Bode's Law?

4 When Neptune was discovered it was found to have an average distance from the Sun of 39.0 astronomical units. How did this fit in with Bode's Law?

5 Does the presence of the asteroids between the orbits of Mars and Jupiter support Bode's Law?

6 Pluto was believed to be a main planet of the Solar System for about 70 years and when it was discovered in 1929 it was found to have an average distance from the Sun of 40.4 astronomical units. How did this fit in with Bode's Law?

7 Why do you think Bode's Law is no longer used?

Index